Mechanical Vibrations
for Engineers

Mechanical Vibrations for Engineers

Michel Lalanne
Patrick Berthier
Johan Der Hagopian
Département de Génie Mécanique Développement
I.N.S.A. de Lyon, France

adapted and translated by
Frederick C. Nelson
College of Engineering, Tufts University, U.S.A.

A Wiley–Interscience Publication

JOHN WILEY AND SONS
Chichester · New York · Brisbane · Toronto · Singapore

Copyright © 1983 by John Wiley & Sons Ltd.

Library of Congress Cataloging in Publication Data:
Lalanne, M.
  Mechanical Vibrations for Engineers.
  'Wiley–Interscience.'
  Bibliography: p.
  Includes index.
  1. Vibration
  I. Berthier, Patrick.   II. Der Hagopian, J.
  III. Nelson, Frederick C.   IV. Title.
TA355.L2413   1984   620.3   83–6908

 ISBN 0 471 90197 0

British Library Cataloguing in Publication Data:
Lalanne, Michel
  Mechanical Vibrations for Engineers.
  1. Vibration.
I. Title   II. Berthier, Patrick
III. Der Hagopian, Johan   IV. Nelson, Frederick C.
V. Mécanique des vibrations linéaires.   *English*
531.11   QA935

ISBN 0 471 90197 0

Printed in Northern Ireland at The Universities Press (Belfast) Ltd

List of Contents

Preface

This book is a revised and extended version of *Mécanique des Vibrations Linéaires* published by Masson in 1980. This version has been translated and adapted for English use by F. C. Nelson.

The book is a basis for the study of linear, mechanical vibrations. It is intended primarily for the use of students and practising mechanical engineers. Its purpose is to provide engineering students and practitioners with:

- an understanding of vibration phenomena and concepts
- the ability to formulate and solve the equations of motion of vibrating systems
- an appreciation of the role and technique of vibration measurement.

The authors have adopted a systematic but practical approach in order to make the book both brief and useful. Special care has been taken to make the presentation easy to read and to understand. It is, however, necessary to have some background in rigid-body mechanics, strength of materials, and matrix calculation.

This book treats the vibrations of structures with a sequence of structural models: first, discrete elements such as springs and masses; then, continuous elements such as beams and plates; and finally, simple structures using finite elements and the method of substructures.

Chapter 1 presents spring–mass systems which have one degree of freedom. In Chapters 2 and 3, these are extended, respectively, to 2 and N degrees of freedom. Chapter 4 presents continuous systems and emphasizes energy methods which then provide an introduction to the widely used finite element method which is described in Chapter 5. Chapter 6 discusses simple measurement devices and introduces modern measurement systems. Chapter 7 presents 12 computer programs written in BASIC and suitable for use with a desk-top computer and graphical display. These programs allow students and practising engineers to consider applications which cannot be considered with hand calculations.

The book also contains over 100 exercises, many of which consider problems of practical interest. All have answers and most have their solutions given in some detail. These exercises are an important part of the book and the reader is encouraged to work as many as possible.

When calculations are performed in an exercise, both with or without the aid of a computer program, at least 8 digits are used for the entire set of calculations. The results are in most cases rounded-off to 4 significant figures. When the results of one exercise are used in another, the rounded-off results of the original exercise are used.

At the end there is an Appendix on Lagrange's equations, a short Bibliography, and an Index.

We wish to thank particularly Mrs. J. Aiello for the careful typing of the manuscript.

1
Single Degree-of-Freedom Systems

The study of single degree-of-freedom systems is a good introduction to the basic phenomena of mechanical vibrations such as resonance, damping, and forced response. It also facilitates an understanding of the behavior of complex systems having a large number of degrees of freedom. In addition, single degree-of-freedom systems can often be used as a convenient first approximation to a real structure. Finally, they are helpful in understanding the behavior of widely used measuring devices such as piezoelectric accelerometers.

1.1 Free Vibration

One degree-of-freedom systems are illustrated by the system shown in Figure 1 for which the motion is assumed to be only vertical. Let x be the displacement of the mass m from the equilibrium position established by the action of gravity; k be the stiffness of the spring; c be the viscous damping coefficient of the damper. The force exerted by the spring on the mass is $-kx$; the force exerted by the damper on the mass is $-cx°$. The time-dependent external force acting on the mass is $F(t)$.

From Newtonian mechanics, the differential equation of motion is

$$mx°° = -cx° - kx + F(t) \tag{1}$$

which for free vibration, that is vibration in the absence of an external force, becomes

$$mx°° + cx° + kx = 0 \tag{2}$$

The solutions of this linear differential equation with constant coefficients

1

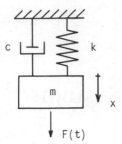

Figure 1 Single degree-of-freedom system with viscous damping

have the exponential form:

$$x = Ae^{rt}$$ (3)

Substituting (3) into (2) gives

$$mr^2 + cr + k = 0$$ (4)

This characteristic equation has two roots r_1 and r_2 given by:

$$r_{1,2} = \frac{1}{2}\left[-\frac{c}{m} \pm \sqrt{\left(\frac{c}{m}\right)^2 - \frac{4k}{m}} \right]$$ (5)

hence the solutions of (2) can be written as:

$$x = A_1 e^{r_1 t} + A_2 e^{r_2 t}$$ (6)

The expression for r_1 and r_2 is best put into a form in which the parameters are much easier to measure. Define the parameters ω and α such as:

$$\omega^2 = \frac{k}{m}$$ (7)

$$\alpha = \frac{c}{c_c}$$ (8)

where ω is the angular frequency in rad/sec and α is the viscous damping ratio. The critical viscous damping coefficient c_c is defined by the vanishing of the discriminant in (5):

$$\left(\frac{c_c}{m}\right)^2 - \frac{4k}{m} = 0$$ (9)

from which

$$c_c = 2\sqrt{km} = 2m\omega$$ (10)

then

$$c = 2m\alpha\omega$$ (11)

and

$$\alpha = \frac{c}{2\sqrt{km}} \tag{12}$$

Using (7) and (11), the expression (5) becomes

$$r_{1,2} = -\alpha\omega \pm \omega\sqrt{\alpha^2 - 1} \tag{13}$$

This clearly shows that the behavior of the system of Figure 1 is completely characterized by the two parameters α and ω. The form of the solution of the differential equation (2) depends on the value of the parameter α.

Case 1: $\alpha < 1$

In practice, this is the most important case. Equation (13) gives

$$r_{1,2} = -\alpha\omega \pm j\omega\sqrt{1 - \alpha^2} \tag{14}$$

with

$$j = \sqrt{-1} \tag{15}$$

and the general solution (6) is written

$$x = A_1 \exp[-\alpha\omega t + j\omega\sqrt{1-\alpha^2}\, t] + A_2 \exp[-\alpha\omega t - j\omega\sqrt{1-\alpha^2}\, t] \tag{16}$$

This can be put into the more convenient form:

$$x = Ae^{-\alpha\omega t} \sin(\omega\sqrt{1-\alpha^2}\, t + \psi)$$
$$= Ae^{-\alpha\omega t} \sin(\omega_d t + \psi) \tag{17}$$

where

$$\omega_d = \omega\sqrt{1-\alpha^2} \tag{18}$$

is the angular frequency of the damped system. A similar analysis shows that ω is the angular frequency of the undamped system. The two constants A and ψ are determined from the two initial conditions: displacement x_0 and velocity $x_0°$ at the initial time t_0.

Also, one can define the frequency f, in hertz (Hz), which is related to the angular frequency ω by

$$\omega = 2\pi f \tag{19}$$

and to the period of oscillation by

$$T = \frac{1}{f} \tag{20}$$

The period is not as widely used in practice as are the frequencies ω and f.

In this first chapter of the text, we are careful to call ω the angular frequency and f the frequency. As has been shown, they are the frequencies with which the undamped system undergoes free vibration. As such, some texts call them the natural frequencies. In subsequent chapters, this distinction in terminology will be dropped and both ω and f will be called the frequency of the system. Which frequency is intended will be clear from the symbol used or the units required.

The experimental determination of α can be achieved by obtaining the logarithmic decrement δ, the natural logarithm of the ratio of two successive maxima of response:

$$\delta = \ln \frac{x_p}{x_{p+1}}$$

$$\approx \ln \frac{\exp[-\alpha \omega t_p]}{\exp[-\alpha \omega (t_p + T)]} \tag{21}$$

This latter expression is an approximation because the points of contact with the exponential envelope curve do not coincide exactly with the maximum response points. From (21),

$$\delta \approx \ln e^{\alpha \omega T}$$

$$\approx \alpha \omega T$$

$$\approx \frac{2\pi\alpha}{\sqrt{1-\alpha^2}} \tag{22}$$

For α small, the usual case in practical situations, this simplifies to

$$\delta \approx 2\pi\alpha \tag{23}$$

The smaller α, the more difficult it will be to measure the ratio of two successive maxima accurately since this ratio will approach unity. It is then better to measure response maxima which are separated by an integral number of periods and

$$\ln \frac{x_p}{x_{p+q}} = q\alpha\omega T$$

$$\approx 2\pi\alpha q \tag{24}$$

hence

$$\alpha \approx \frac{1}{2\pi q} \ln \frac{x_p}{x_{p+q}} \tag{25}$$

Case 2: $\alpha = 1$
This case is seldom encountered in mechanical systems. The characteristic equation has a double root:

$$r_{1,2} = -\omega \tag{26}$$

The response is aperiodic with critical damping and is given by

$$x = A_1 e^{-\omega t} + A_2 t e^{-\omega t}$$

$$= e^{-\omega t}(A_1 + A_2 t) \tag{27}$$

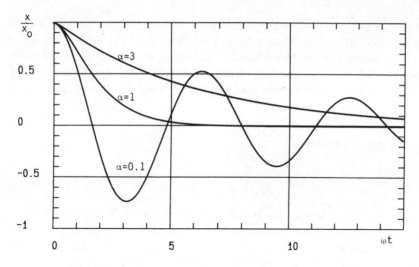

Figure 2   Free motion of the system of Figure 1 for various amounts of damping

Case 3: $\alpha > 1$
This case is also rare in mechanical systems. The solution has the form

$$x = A_1 \exp\left[-\alpha\omega t + \omega\sqrt{\alpha^2 - 1}\ t\right] + A_2 \exp\left[-\alpha\omega t - \omega\sqrt{\alpha^2 - 1}\ t\right] \quad (28)$$

Figure 2 shows the free vibration of a system with initial conditions $x = x_0$, $x° = 0$ for $t_0 = 0$ and for damping ratios of $\alpha = 0.1$, 1, and 3. The calculations give, for:

$$\alpha = 0.1: \quad x = 1.005 x_0 e^{-0.1\omega t} \sin (0.995\omega t + 1.471)$$
$$\alpha = 1: \quad x = x_0(1 + \omega t)e^{-\omega t} \quad (29)$$
$$\alpha = 3: \quad x = 1.030 x_0 e^{-0.1716\omega t} - 0.030 x_0 e^{-5.828\omega t}$$

In these three types of free response, x tends toward zero and this fact will justify the subsequent decision to ignore the transient part of the forced response for large time. Also notice that the case $\alpha = 1$ returns the system to rest in the minimum time.

1.2   Forced Vibration

The general solution of equation (1) is equal to the sum of the solution to the homogeneous equation, (2), and a particular solution of (1). In the most frequent case of $\alpha < 1$, one has

$$x = Ae^{-\alpha\omega t} \sin (\omega\sqrt{1 - \alpha^2}\ t + \psi)$$
$$+ \text{a particular solution of equation (1)} \quad (30)$$

Frequently only the steady-state motion is of interest; that is, the system motion existing after a sufficient length of time that the initial transient motion associated with the free vibration has become negligible. Three cases of excitation will be considered: harmonic; periodic; and general function of time. The solutions to harmonic and periodic excitation will be limited to steady-state motion.

1.2.1 Harmonic excitation

Let

$$F(t) = F \sin \Omega t \tag{31}$$

where F is the amplitude of the exciting force and Ω is the angular forcing frequency in rad/sec. The symbol Ω will be used for the angular forcing frequency to distinguish it from ω, the angular frequency of the system. In subsequent chapters, Ω will simply be called the forcing frequency.

Equation (1) becomes

$$mx^{\circ\circ} + cx^{\circ} + kx = F \sin \Omega t \tag{32}$$

The steady-state solution has the form

$$x = \sin (\Omega t - \phi) \tag{33}$$

where X is the amplitude of the steady-state solution and ϕ is the phase angle. Substituting (33) in (32) gives

$$(k - m\Omega^2)X \sin (\Omega t - \phi) + c\Omega X \cos (\Omega t - \phi) = F \sin \Omega t \tag{34}$$

which can be written as

$$[c\Omega \cos \phi - (k - m\Omega^2) \sin \phi]X \cos \Omega t$$
$$+ [(k - m\Omega^2)X \cos \phi + c\Omega X \sin \phi - F] \sin \Omega t = 0 \tag{35}$$

Equation (35) holds for any time t, hence

$$c\Omega \cos \phi - (k - m\Omega^2) \sin \phi = 0 \tag{36}$$

$$[(k - m\Omega^2) \cos \phi + c\Omega \sin \phi]X - F = 0 \tag{37}$$

Equation (36) permits $\sin \phi$ to be obtained in terms of $\cos \phi$:

$$\sin \phi = \frac{c\Omega \cos \phi}{k - m\Omega^2} \tag{38}$$

On substituting this expression into (37) one obtains

$$\cos \phi = \frac{F}{X} \frac{k - m\Omega^2}{(k - m\Omega^2)^2 + c^2\Omega^2} \tag{39}$$

in which F, X, and $(k - m\Omega^2)^2 + c^2\Omega^2$ are positive quantities.

Using (38) and (39), it can be shown that:

$$\text{for} \quad \Omega < \sqrt{\frac{k}{m}} = \omega$$

$$\sin \phi > 0$$
$$\cos \phi > 0$$

$$\text{hence} \quad 0 < \phi < \frac{\pi}{2} \qquad (40)$$

$$\text{for} \quad \Omega > \sqrt{\frac{k}{m}} = \omega$$

$$\sin \phi > 0$$
$$\cos \phi < 0$$

$$\text{hence} \quad \frac{\pi}{2} < \phi < \pi \qquad (41)$$

Since $0 < \phi < \pi$ the phase can be uniquely defined by its tangent taken from (38):

$$\tan \phi = \frac{c\Omega}{k - m\Omega^2}$$
$$= \frac{2\alpha(\Omega/\omega)}{1 - (\Omega/\omega)^2} \qquad (42)$$

Using the identity

$$\cos^2 \phi + \sin^2 \phi = 1$$

and (38) and (39),

$$X = \frac{F}{\sqrt{(k - m\Omega^2)^2 + c^2\Omega^2}} = \frac{X_{st}}{\sqrt{[1 - (\Omega/\omega)^2]^2 + [2\alpha(\Omega/\omega)]^2}} \qquad (43)$$

with

$$X_{st} = \frac{F}{k} \qquad (44)$$

which is the displacement of the system subjected to a static force F.
    It can be shown that the amplitude X has a maximum for

$$\frac{\Omega}{\omega} = \sqrt{1 - 2\alpha^2} \qquad (45)$$

and that the corresponding value of X is

$$X_r = \frac{X_{st}}{2\alpha\sqrt{1-\alpha^2}} \tag{46}$$

At this maximum, the tangent of the phase angle is

$$\tan\phi = \frac{\sqrt{1-2\alpha^2}}{\alpha} \tag{47}$$

This is the definition of amplitude resonance. Note that if $\alpha > 1/\sqrt{2}$, the maximum amplitude occurs at $\Omega = 0$.

If, on the other hand, one takes

$$\frac{\Omega}{\omega} = 1$$

$\tan\phi$ becomes infinite at $\phi = \pi/2$ and $X_r = X_{st}/2\alpha$. This is the definition of phase resonance.

For most practical systems the damping is small, i.e. $\alpha < 0.1$, and (45), (46), and (47) become

$$\frac{\Omega}{\omega} \simeq 1-\alpha^2 \simeq 1 \tag{48}$$

$$X_r = \frac{X_{st}}{2\alpha}\left(1+\frac{\alpha^2}{2}\right) \tag{49}$$

$$\simeq \frac{X_{st}}{2\alpha} = Q \cdot X_{st} \tag{50}$$

$$\tan\phi \simeq \frac{1}{\alpha}$$

where Q is the so-called Q-factor. This definition of Q can be used for any form of damping.

Thus for small damping the angular forcing frequency associated with amplitude resonance and that associated with phase resonance are essentially equal and it is not necessary to distinguish between them. Under these restrictions, resonance occurs when Ω matches the frequency of the system for free vibration, ω. For this reason, ω will often be referred to as the resonant frequency.

At resonance, the forces in the system elements can be very large if the damping is small. In the spring the force amplitude is

$$F_r = kX_r$$

$$= \frac{kX_{st}}{2\alpha} = \frac{F}{2\alpha} = Q \cdot F \tag{51}$$

which can become very large because of the term $1/2\alpha$. This, of course, is the reason why the determination of resonance is important in structures.

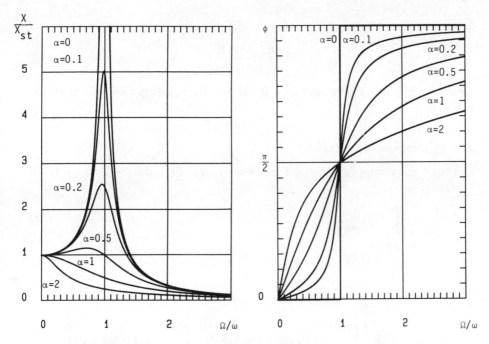

Figure 3   Steady-state amplitude and phase response of the system of Figure 1 for a sinusoidal forcing function

The curves of Figure 3 show X/X_{st} and ϕ as a function of Ω/ω for several values of the damping factor α.

For the case of damping so small that we may take $\alpha = 0$, equations (42) and (43) give:

$$\Omega < \omega: \quad \phi = 0 \quad \text{and} \quad x = \frac{X_{st} \sin \Omega t}{1 - (\Omega/\omega)^2} \tag{52}$$

$$\Omega = \omega: \quad \phi = \frac{\pi}{2} \quad \text{and} \quad x \to \infty$$

$$\Omega > \omega: \quad \phi = \pi \quad \text{and} \quad x = \frac{X_{st} \sin (\Omega t - \pi)}{(\Omega/\omega)^2 - 1} \tag{53}$$

These are the equations for the steady-state response of the undamped system.

The results (52) and (53) can also be obtained by considering the solution of

$$m x^{\infty} + kx = F \sin \Omega t \tag{54}$$

in the form

$$x = X \sin \Omega t \tag{55}$$

from which

$$x = \frac{F}{k} \frac{\sin \Omega t}{1 - (\Omega/\omega)^2} \qquad (56)$$

which is the same expression as (52). If $\omega < \Omega$, equation (56) will have the same form as (53).

Energy dissipation per cycle

The energy supplied by the external force $F(t)$ during one cycle of vibration is

$$E = \int_0^T F(t) \frac{dx}{dt} dt \qquad (57)$$

$$= \int_0^{2\pi/\Omega} \Omega X F(t) \cos (\Omega t - \phi) \, dt$$

$$= \Omega X F \int_0^{2\pi/\Omega} \sin \Omega t \cos (\Omega t - \phi) \, dt$$

$$= \Omega X F \left[\int_0^{2\pi/\Omega} \sin \Omega t \cos \Omega t \cos \phi \, dt + \int_0^{2\pi/\Omega} \sin^2 \Omega t \sin \phi \, dt \right]$$

$$= \pi X F \sin \phi \qquad (58)$$

Using (38) and (39), one has

$$F \sin \phi = \frac{c \Omega \cos \phi}{k - m\Omega^2} F$$

$$= \frac{c \Omega F^2}{X[(k - m\Omega^2)^2 + c^2 \Omega^2]} \qquad (59)$$

$$= c \Omega X$$

and combining (58) and (59), the expression for energy dissipation in one cycle of motion becomes

$$E = \pi c \Omega X^2 \qquad (60)$$

One could also obtain this result by considering the energy dissipated per cycle in the damper:

$$E = \int_0^{2\pi/\Omega} c x^\circ \frac{dx}{dt} dt \qquad (61)$$

Bandwidth

For sufficiently small values of damping, the bandwidth is the frequency interval $(f_2 - f_1)$ situated around the resonance frequency such that the amplitude of response at frequencies f_1 and f_2 is equal to the resonant amplitude divided by $\sqrt{2}$. Equation (60) shows that the energies dissipated at the frequencies f_1 and f_2 are then equal to one-half of the energy dissipated at resonance. For this reason, the above bandwidth is sometimes called the half-power bandwidth. Then if α is small,

$$X(\Omega_1) = X(\Omega_2) = \frac{X_r}{\sqrt{2}} = \frac{1}{\sqrt{2}} \frac{X_{st}}{2\alpha} \tag{62}$$

Ω_1 and Ω_2 must be solutions of the following equation:

$$\frac{1}{2\alpha\sqrt{2}} = \frac{1}{\sqrt{[1-(\Omega/\omega)^2]^2 + [2\alpha\Omega/\omega]^2}} \tag{63}$$

Rewriting equation (63) gives

$$\left(\frac{\Omega}{\omega}\right)^4 + (4\alpha^2 - 2)\left(\frac{\Omega}{\omega}\right)^2 + 1 - 8\alpha^2 = 0 \tag{64}$$

hence

$$\left(\frac{\Omega}{\omega}\right)^2 = 1 - 2\alpha^2 \pm \sqrt{(1-2\alpha^2)^2 - (1-8\alpha^2)}$$

$$= 1 - 2\alpha^2 \pm 2\alpha\sqrt{1+\alpha^2}$$

$$\simeq 1 \pm 2\alpha \tag{65}$$

then

$$\left(\frac{\Omega_2}{\omega}\right)^2 - \left(\frac{\Omega_1}{\omega}\right)^2 \simeq 4\alpha \tag{66}$$

or:

$$\left(\frac{\Omega_2}{\omega}\right)^2 - \left(\frac{\Omega_1}{\omega}\right)^2 = \frac{\Omega_2 - \Omega_1}{\omega} \frac{\Omega_2 + \Omega_1}{\omega} \tag{67}$$

$$\simeq \frac{\Delta\Omega}{\omega^2} 2\omega \tag{68}$$

From the relations (66) and (68), one finds

$$\frac{\Delta\Omega}{\omega} = \frac{\Delta f}{f} = 2\alpha = \frac{1}{Q} \tag{69}$$

with $\Delta f = f_2 - f_1$, the bandwidth, and f the resonant frequency. A very useful way of determining α is to measure the bandwidth. Note that

$$20 \log_{10} \frac{X(\Omega_1)}{X_r} = 20 \log_{10} \frac{X(\Omega_2)}{X_r} \simeq -3 \text{ dB} \tag{70}$$

which justifies the frequently used technique of determining the half power bandwidth by locating the frequencies on either side of resonance for which the response has decreased by 3 decibels (dB).

1.2.2 Periodic excitation

Consider next an excitation force which can be developed in a Fourier series. If the forcing frequency is given by $\Omega = 2\pi/T$, this series has the form

$$F(t) = \frac{a_0}{2} + \sum_{p=1}^{\infty} (a_p \cos p\Omega t + b_p \sin p\Omega t) \tag{71}$$

with

$$a_p = \frac{2}{T} \int_0^T F(t) \cos p\Omega t \, dt \qquad p = 0, 1, 2, \ldots \tag{72}$$

$$b_p = \frac{2}{T} \int_0^T F(t) \sin p\Omega t \, dt \qquad p = 1, 2, \ldots \tag{73}$$

In steady-state motion, the response to each harmonic component is calculated separately and these responses are then added to obtain the complete solution.

Consider, for example, a spring–mass system with small damping under the

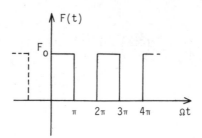

Figure 4   Periodic square-wave
forcing function

action of the force shown in Figure 4. Using equations (71), (72), and (73):

$$F(t) = \frac{F_0}{2} + \frac{2F_0}{\pi} \sum_{p=1,3,\ldots}^{\infty} \frac{\sin p\Omega t}{p} \tag{74}$$

The system equation is then

$$m\ddot{x} + kx = \frac{F_0}{2} + \frac{2F_0}{\pi} \sum_{p=1,3,\ldots}^{\infty} \frac{\sin p\Omega t}{p} \tag{75}$$

The solutions to (75) for each term of $F(t)$ given in (74) is obtained in the same manner as used for (56). It follows that

$$x = \frac{F_0}{2k} + \frac{2F_0}{\pi} \sum_{p=1,3,\ldots}^{\infty} \frac{\sin p\Omega t}{p[k - m(p\Omega)^2]} \tag{76}$$

Resonance will occur whenever

$$k - m(p\Omega)^2 = 0 \tag{77}$$

which is equivalent to

$$\Omega = \frac{1}{p}\sqrt{\frac{k}{m}} = \frac{\omega}{p} \tag{78}$$

1.2.3 General function of time

In this case, the system response is required starting from the initial time $t_0 = 0$. In practice, a numerical step-by-step method is usually used (see Chapter 3). However, one can also solve this problem with the Laplace transform applied to equation (1). This method will be briefly reviewed without a discussion on the question of convergence.

For a function $f(t)$ defined for $t > 0$ and zero otherwise, the Laplace transform, $\mathcal{L}(p)$, is given by:

$$L[f(t)] = \mathcal{L}(p) = \int_0^{\infty} e^{-pt} f(t)\, \mathrm{d}t \tag{79}$$

The Laplace transform of the first and second derivatives of $f(t)$ are obtained from (79). If $f(0)$ and $f°(0)$ are respectively the values of $f(t)$ and of its first derivative at $t = 0$, integration by parts gives

$$L\left[\frac{\mathrm{d}f(t)}{\mathrm{d}t}\right] = -f(0) + p\mathcal{L}(p) \tag{80}$$

$$L\left[\frac{\mathrm{d}^2 f(t)}{\mathrm{d}t^2}\right] = -f°(0) - pf(0) + p^2\mathcal{L}(p) \tag{81}$$

From equations (1), (80), and (81), the Laplace transform of $x(t)$ is found to be $\chi(p)$, where

$$\chi(p) = \frac{\mathcal{L}(p)}{m(p^2 + 2\alpha\omega p + \omega^2)} + \frac{(p + 2\alpha\omega)x(0)}{p^2 + 2\alpha\omega p + \omega^2} + \frac{x°(0)}{p^2 + 2\alpha\omega p + \omega^2} \tag{82}$$

and $\mathcal{L}(p)$ is the Laplace transform of the exciting force. The last two terms of (82) are well-known Laplace transforms of simple functions and hence can be easily inverted. But to invert the first term, Borel's theorem must be used. This theorem shows that if $\mathcal{L}_1(p)$, $\mathcal{L}_2(p)$ are the transforms of $f_1(t)$, $f_2(t)$, the transform

$$\mathcal{L}(p) = \mathcal{L}_1(p) \cdot \mathcal{L}_2(p) \tag{83}$$

originates from

$$f(t) = \int_0^t f_1(\tau) \cdot f_2(t-\tau) \, d\tau \tag{84}$$

The inverse of (82) is then

$$x(t) = \frac{1}{m\omega_d} \int_0^t F(\tau) e^{-\alpha\omega(t-\tau)} \sin \omega_d(t-\tau) \, d\tau$$

$$+ x(0)e^{-\alpha\omega t} \left(\cos \omega_d t + \frac{\alpha}{(1-\alpha^2)^{1/2}} \sin \omega_d t \right) + \frac{x°(0)}{\omega_d} e^{-\alpha\omega t} \sin \omega_d t \tag{85}$$

where the terms containing $x(0)$ and $x°(0)$ correspond to the transient response. This form shows very clearly that the transient response is necessary to satisfy the initial conditions at $t = 0$ but as time grows large it decays to zero, leaving only the steady-state solution. Also notice that one must carefully distinguish between ω and ω_d in (85).

Thus the steady-state motion of a damped system is given by

$$x(t) = \frac{1}{m\omega_d} \int_0^t F(\tau) e^{-\alpha\omega(t-\tau)} \sin \omega_d(t-\tau) \, d\tau \tag{86}$$

and for an undamped or very lightly damped system by

$$x(t) = \frac{1}{m\omega} \int_0^t F(\tau) \sin \omega(t-\tau) \, d\tau \tag{87}$$

1.3 Damping in Real Systems

Actual systems always have some damping but rarely is this damping viscous. Among the most common forms of damping are structural damping and Coulomb damping. Structural damping is a material characteristic whose value can be strongly dependent on both temperature and forcing frequency. Coulomb damping arises from the relative motion between dry surfaces in contact; it is quite difficult to quantify this phenomenon because it depends on so many parameters.

An equivalent viscous damping coefficient can be defined for the case of harmonic excitation by using the previous expression for energy dissipated per cycle (60).

For structural damping it has been observed that the energy dissipated per cycle has the form

$$E = aX^2 \tag{88}$$

over a limited range of frequency and temperature. X is the displacement amplitude and a is a constant of proportionality. The coefficient of equivalent viscous damping is found from (60) and (88):

$$aX^2 = \pi c_{eq} \Omega X^2 \tag{89}$$

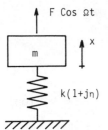

F Cos Ωt

x

m

k(1+jn)

Figure 5 Single degree-of-freedom system with structural damping

hence

$$c_{eq} = \frac{a}{\pi\Omega} \tag{90}$$

The calculation of systems with structural damping subjected to harmonic excitation is more conveniently achieved with the use of complex notation. A single degree-of-freedom system with structural damping and excited by the force $F \cos \Omega t$ is shown in Figure 5. It has the equation

$$mx^{\infty} + \frac{a}{\pi\Omega} x^{\circ} + kx = F \cos \Omega t \tag{91}$$

In complex notation, this becomes

$$mz^{\infty} + \frac{a}{\pi\Omega} z^{\circ} + kz = Fe^{j\Omega t} \tag{92}$$

where $x = \text{Re}[z]$, the real part of the complex quantity z. Solutions are sought in the form

$$z = Ze^{j\Omega t} \tag{93}$$

which, when substituted in (92), gives

$$(k - m\Omega^2)Z + j\frac{a}{\pi} Z = F \tag{94}$$

Equation (94) is conveniently written as:

$$-m\Omega^2 Z + k(1 + j\eta)Z = F \tag{95}$$

with

$$\eta = \frac{a}{\pi k} \qquad \text{the structural damping factor} \tag{96}$$

and, in addition, one can define

$$k^* = k(1 + j\eta) \qquad \text{the complex stiffness} \tag{97}$$

The term η in equations (96) and (97) is often referred to as the loss factor.

Equation (95) gives

$$Z = \frac{F}{k - m\Omega^2 + j\eta k} \tag{98}$$

which can be put into the form

$$Z = |Z|e^{-i\phi} \tag{99}$$

with

$$|Z| = \frac{F}{\sqrt{(k - m\Omega^2)^2 + \eta^2 k^2}} \tag{100}$$

and where

$$\sin \phi = \frac{\eta k}{\sqrt{(k - m\Omega^2)^2 + \eta^2 k^2}}$$
$$\cos \phi = \frac{k - m\Omega^2}{\sqrt{(k - m\Omega^2)^2 + \eta^2 k^2}} \tag{101}$$

Since

$$x = \mathrm{Re}\,[Ze^{i\Omega t}] \tag{102}$$

one has

$$x = \frac{F}{\sqrt{(k - m\Omega^2)^2 + \eta^2 k^2}} \cos(\Omega t - \phi) \tag{103}$$

with

$$\tan \phi = \frac{\eta k}{k - m\Omega^2} \qquad 0 < \phi < \pi \tag{104}$$

Equations (103) and (104) are more conveniently written in the form

$$x = \frac{F/k}{\sqrt{[1 - (\Omega/\omega)^2]^2 + \eta^2}} \cos(\Omega t - \phi) \tag{105}$$

$$\tan \phi = \frac{\eta}{1 - (\Omega/\omega)^2} \tag{106}$$

The determination of η is easy because

$$\eta = \frac{\Delta f}{f} \tag{107}$$

where Δf is the half-power bandwidth. This then provides a means of measuring damping during steady-state vibration (see exercise 15).

It is important to note that the use of complex notation results in a major reduction in the algebra required to solve the equation of motion of a harmonically forced oscillator: compare the number of steps required to get (43) with the number of steps to get (95). For this reason, complex notation will be used in subsequent analysis.

1.4 Rayleigh's Method

This method was proposed by Rayleigh to obtain a close approximation to the lowest frequency of free vibration for an undamped system. In this chapter, Rayleigh's method will be applied to single degree-of-freedom systems only.

The method proceeds as follows: from a reasonable hypothesis about the motion of the system, calculate the approximate kinetic and strain energies and then use the theorem of conservation of mechanical energy or the equations of Lagrange to obtain the approximate frequency.

Let us apply the method to estimate the effect of the mass of the spring on the dynamic behavior of a spring–mass system, as shown in Figure 6, where: $m_s = \rho L$, mass of the spring; L, spring length; ρ, mass per unit length of the spring.

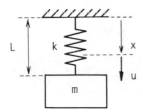

Figure 6   Influence of spring mass on the frequency of a spring–mass system

The motion of the system is assumed to be separable; that is, if $u(x, t)$ is the displacement of a point on the spring located at a distance x from the fixed end, then it is assumed that $u(x, t)$ can be separated into the product

$$u(x, t) = \phi(x)p(t) \tag{108}$$

By analogy with the static case, a reasonable hypothesis for $\phi(x)$ is

$$\phi(x) = ax \tag{109}$$

where a is a constant of proportionality. The expression for kinetic energy is

$$T = T_{\text{mass}} + T_{\text{spring}} \tag{110}$$

$$= \tfrac{1}{2}ma^2L^2p^{\circ 2}(t) + \frac{1}{2}\int_0^L \rho a^2x^2\,dx\,p^{\circ 2}(t)$$

$$= \frac{1}{2}\left(m + \frac{m_s}{3}\right)a^2L^2p^{\circ 2}(t) \tag{111}$$

and the expression for strain energy is

$$U = \tfrac{1}{2}ka^2L^2p^2(t) \tag{112}$$

Application of Lagrange's equation gives

$$\left(m + \frac{m_s}{3}\right)p^\infty(t) + kp(t) = 0 \tag{113}$$

and then the approximate angular frequency is

$$\omega = \sqrt{\frac{k}{m + m_s/3}} \tag{114}$$

As stated above, this result can also be obtained by writing that the sum of (111) and (112) is constant in time.

1.5 Applications

The two applications described below are important because they are first-order models for systems such as vehicle suspensions, accelerometers, and vibration isolators. Only the expressions for response amplitudes are developed since the phase responses are usually of less interest.

1.5.1 System on a moving foundation

The system represented in Figure 7 models a machine mounted to a foundation by a spring and viscous damper. The foundation has a displace-

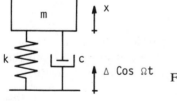

Figure 7 Single degree-of-freedom system on a moving foundation

ment of the form

$$\delta = \Delta \cos \Omega t \tag{115}$$

and it is desired to keep the machine motion, that is the motion of the mass, to the smallest possible value. This situation also arises for a vehicle going over a rough road or for a container of delicate electronics attached to a vibrating surface.

The movement of the mass can be deduced from the equation of motion:

$$mx^\infty = k(\delta - x) + c(\delta^\circ - x^\circ) \tag{116}$$

Substituting (115) into (116) gives

$$mx^{\infty} + cx^{\circ} + kx = \Delta(k \cos \Omega t - c\Omega \sin \Omega t) \qquad (117)$$

In order to use complex notation, we associate (118) with (117):

$$j(my^{\infty} + cy^{\circ} + ky) = j\Delta(k \sin \Omega t + c\Omega \cos \Omega t) \qquad (118)$$

and obtain the equation

$$mz^{\infty} + cz^{\circ} + kz = \Delta(k + jc\Omega)e^{j\Omega t} \qquad (119)$$

The real part of z is x. The amplitude of the mass displacement X is

$$X = \Delta \sqrt{\frac{k^2 + c^2\Omega^2}{(k - m\Omega^2)^2 + c^2\Omega^2}} \qquad (120)$$

$$= \Delta \sqrt{\frac{1 + [2\alpha(\Omega/\omega)]^2}{[1 - (\Omega/\omega)^2]^2 + [2\alpha(\Omega/\omega)]^2}} \qquad (121)$$

The ratio X/Δ is plotted in Figure 8 as a function of Ω/ω with α as a parameter. In order to have a small motion of the mass, that is, good isolation, it is required that $\Omega/\omega \gg 1$. In other words, the resonant frequency of the system, ω, must be as low as possible. In practice, this is limited by the

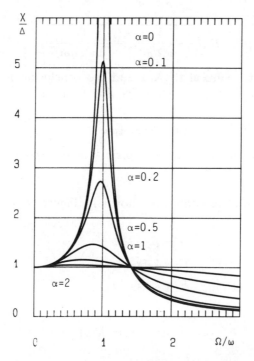

Figure 8 Transmissibility curves for a single degree-of-freedom system

increasing static displacement of the mass due to gravity:

$$kx_{st} = mg \tag{122}$$

1.5.2 Transmissibility

Now let the mass of the system just considered be subjected to a force $F \sin \Omega t$ and require that the force transmitted to the foundation, which is now fixed, be as small as possible.

The force transmitted is

$$F_t = kx + cx^\circ \tag{123}$$

or if

$$x = X \sin (\Omega t - \phi) \tag{124}$$

$$F_t = X[k \sin (\Omega t - \phi) + c\Omega \cos (\Omega t - \phi)] \tag{125}$$

The amplitude of F_t can be shown to be

$$|F_t| = X\sqrt{k^2 + c^2\Omega^2}$$

$$= kX \sqrt{1 + [2\alpha(\Omega/\omega)]^2} \tag{126}$$

and after using (43),

$$|F_t| = F\sqrt{\frac{1 + [2\alpha(\Omega/\omega)]^2}{[1 - (\Omega/\omega)^2]^2 + [2\alpha(\Omega/\omega)]^2}} \tag{127}$$

The ratio $|F_t|/F$ is identical to X/Δ and the conclusion is therefore: to limit the transmitted force, it is necessary that $\Omega/\omega \gg 1$.

1.6 Exercises

Exercise 1: *Calculate the equivalent stiffness k of two springs of stiffness k_1 and k_2 which are in parallel.*

The forces and displacements are shown in Figure 9 in their positive directions. Then

$$F = F_1 + F_2$$
$$F_1 = k_1(x_2 - x_1)$$
$$F_2 = k_2(x_2 - x_1)$$

hence

$$F = (k_1 + k_2)(x_2 - x_1) = k(x_2 - x_1)$$

and

$$k = k_1 + k_2$$

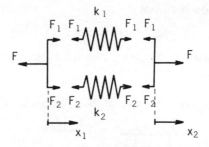

Figure 9

Exercise 2: *Repeat exercise 1 for two springs in series.*

The answer is

$$\frac{1}{k} = \frac{1}{k_1} + \frac{1}{k_2}$$

Exercise 3: *The influence of gravity. Consider the vertical spring–damper–mass system shown in Figure 10. Derive the equation of motion using as origin for X the unextended length of the spring.*

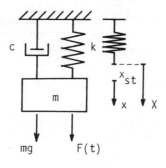

Figure 10

Newton's law gives

$$mX^{\circ\circ} + cX^{\circ} + kX = F(t) + mg$$

In introducing

$$X = x_{st} + x$$

where

$$x_{st} = \frac{mg}{k}$$

the equation of motion becomes

$$mx^{\infty} + cx^{\circ} + kx = F(t)$$

The influence of gravity is therefore just a change of origin and does not affect the dynamic response.

Exercise 4: *Find the free vibration of the system shown in Figure 11 for*

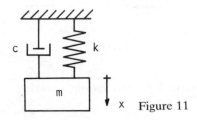

x   Figure 11

$\alpha < 1.0$ *with initial condition* $x(0) = x_0$ *and* $x^{\circ}(0) = x_0^{\circ}$.

The displacement $x(t)$ has the form:

$$x = Ae^{-\alpha\omega t} \sin(\omega\sqrt{1-\alpha^2}\, t + \psi)$$

Applying the initial conditions gives

$$x_0 = A \sin \psi$$

$$x_0^{\circ} = -\alpha\omega A \sin \psi + A\omega\sqrt{1-\alpha^2} \cos \psi$$

$$\tan \psi = \frac{x_0\omega\sqrt{1-\alpha^2}}{x_0^{\circ} + \alpha\omega x_0}$$

When ψ is known, one can find $\sin \psi$ and then A.

Exercise 5: *Consider the same system as in exercise 4. Find the free vibration of the system for* $\alpha = 0.05$ *and the initial conditions* $x(0) = 0$, $x^{\circ}(0) = x_0^{\circ}$.

The answer is

$$x(t) = \frac{1.00125}{\omega} x_0^{\circ} e^{-0.05\omega t} \sin 0.99875\, \omega t$$

Exercise 6: *The base of the system shown in Figure 12 has a prescribed displacement* $\delta = \Delta \sin \Omega t$. *Calculate the steady-state response and the base force necessary to impose this displacement.*

The equation of motion is

$$mx^{\infty} + kx = k\Delta \sin \Omega t$$

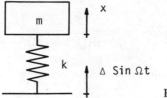

Figure 12

If the motion of the mass has the form

$$x = X \sin \Omega t$$

the response is

$$x = \frac{\Delta}{1 - (\Omega/\omega)^2} \sin \Omega t$$

The force necessary to produce this displacement is

$$F = k(\delta - x)$$
$$= -\frac{k\Delta(\Omega/\omega)^2}{1 - (\Omega/\omega)^2} \sin \Omega t$$

At resonance of the undamped system, this force becomes infinite. In practical testing situations, since it is impossible to produce a force of this magnitude, it is necessary to significantly reduce the displacement amplitude whenever a system becomes resonant.

Exercise 7: *Consider the same system as shown in exercise 4. Find the steady-state motion of the system subjected to $F(t) = F \sin \Omega t$ by using complex notation.*

An equation

$$my^{\infty} + cy^{\circ} + ky = F \cos \Omega t$$

can be combined with the standard equation (32):

$$mx^{\infty} + cx^{\circ} + kx = F \sin \Omega t$$

so that x is the imaginary part of z, which is the solution of

$$mz^{\infty} + cz^{\circ} + kz = Fe^{j\Omega t}$$

where

$$z = Ze^{j\Omega t}$$
$$= |Z| e^{j(\Omega t - \phi)}$$

It can be shown that

$$|Z| = \frac{F}{\sqrt{(k - m\Omega^2)^2 + c^2\Omega^2}}$$

$$\cos \phi = \frac{k - m\Omega^2}{\sqrt{(k - m\Omega^2)^2 + c^2\Omega^2}}$$

$$\sin \phi = \frac{c\Omega}{\sqrt{(k - m\Omega^2)^2 + c^2\Omega^2}}$$

Equation (42) for $\tan \phi$ can then be obtained and it can be seen that $0 < \phi < \pi$. Finally,

$$x = \text{Im}\,[z(t)]$$
$$= \text{Im}\,[|Z|\, e^{j(\Omega t - \phi)}]$$
$$= \frac{F}{\sqrt{(k - m\Omega^2)^2 + (c\Omega)^2}} \sin (\Omega t - \phi)$$

Exercise 8: *Write the equation of motion in terms of θ for the system shown in Figure 13 for the case of small θ. The beam ADB is rigid and is pinned at*

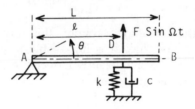

Figure 13

point A and free at point B. The mass moment of inertia of the beam with respect to A is I_A. Find the value of the damping factor α if the values of the various system parameters are as follows:

	I_A	k	c	l	L
SI	0.065 m² · kg	3×10^4 N/m	1 Ns/m	0.4 m	0.5 m
English units	0.048 ft² slug	2056 lbf/ft	0.0685 lbf · sec/ft	1.31 ft	1.64 ft

The equation of motion in SI units is

$$0.065\ddot{\theta} + 0.16\dot{\theta} + 4800\theta = 0.4F \sin \Omega t$$

and therefore

$$\alpha = \frac{0.16}{2\sqrt{0.065 \times 4800}} = 0.45\%$$

The equation of motion in English units is

$$0.048\theta^{\infty} + 0.118\theta^{\circ} + 35280\theta = 1.31 F \sin \Omega t$$

Exercise 9: *For the system shown in exercise 8, calculate the steady-state response in terms of the vertical displacement x(t) at point B. Draw X/F as a function of Ω using both log-log and linear-linear scales. Discuss the approximations which can be obtained for frequencies much less than the resonant frequency and for frequencies much more than the resonant frequency.*

The equation of motion (in SI units) is

$$0.13x^{\infty} + 0.32x^{\circ} + 9600x = 0.4F \sin \Omega t$$

which is a specific form of:

$$m^* x^{\infty} + c^* x^{\circ} + k^* x = F^* \sin \Omega t$$

The solution is

$$x = X \sin (\Omega t - \phi)$$

with

$$X = \frac{F^*}{\sqrt{(k^* - m^* \Omega^2)^2 + (c^* \Omega)^2}}$$

At low frequency (i.e. $\Omega \ll \sqrt{k/m}$):

$$X \simeq \frac{F^*}{k^*}$$

At high frequency (i.e. $\Omega \gg \sqrt{k/m}$):

$$X \simeq \frac{F^*}{m^* \Omega^2}$$

In this latter case, on the logarithm scales:

$$\log \frac{X}{F^*} = -\log (m^* \Omega^2) = -\log m^* - 2 \log \Omega$$

which corresponds to a straight line of slope -2.

The graphs are shown in Figure 14. From this exercise one is able to see that logarithm scales are more appropriate for showing system response at low and high frequencies than are linear scales.

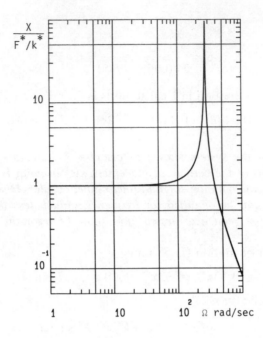

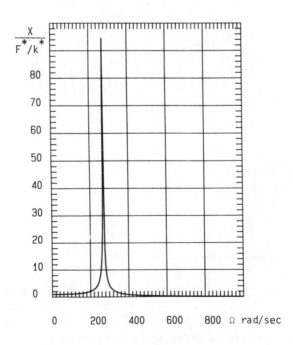

Figure 14

Exercise 10: *The mass shown in Figure 15 is subjected to a force step of magnitude F at t = 0. The initial conditions are x(0) = 0, x°(0) = 0. Calculate*

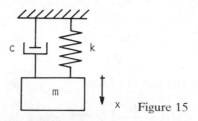

Figure 15

x(t) for $\alpha = 0$ and for $\alpha \neq 0$.

The equation is

$$mx^{\circ\circ} + cx^{\circ} + kx = F \qquad (t > 0)$$

For $\alpha = 0$, the general solution is

$$x = A \sin \omega t + B \cos \omega t + \frac{F}{k}$$

and after applying the initial conditions this becomes

$$x = \frac{F}{k}(1 - \cos \omega t)$$

For $\alpha \neq 0$, the general solution is

$$x = e^{-\alpha\omega t}(A \sin \omega\sqrt{1-\alpha^2}\, t + B \cos \omega \sqrt{1-\alpha^2}\, t) + \frac{F}{k}$$

where

$$A = \frac{-\alpha F}{k\sqrt{1-\alpha^2}}, \qquad B = -\frac{F}{k}$$

Hence

$$x = -\frac{F}{k}e^{-\alpha\omega t}\left(\frac{\alpha}{\sqrt{1-\alpha^2}} \sin \omega\sqrt{1-\alpha^2}\, t + \cos \omega\sqrt{1-\alpha^2}\, t\right) + \frac{F}{k}$$

Exercise 11: *Consider the same system as in exercise 10 for the case $\alpha \neq 0$ but now consider the force step of magnitude F to be a general function of time. Calculate x(t) for the same initial conditions.*

Due to the initial conditions, equation (86) is used:

$$x(t) = \frac{1}{m\omega_d} \int_0^t F(\tau)e^{-\alpha\omega(t-\tau)} \sin \omega_d(t - \tau)\, d\tau$$

In this case

$$F(\tau) = F$$

and, if one lets

$$t - \tau = u$$
$$d\tau = -du$$

$x(t)$ can be written in the simpler form:

$$x(t) = \frac{F}{m\omega_d} \int_0^t e^{-\alpha\omega u} \sin \omega_d u \, du$$

from which:

$$x(t) = \frac{F}{m\omega\sqrt{1-\alpha^2}} \int_0^t e^{-\alpha\omega u} \sin \omega_d u \, du$$

or since:

$$\int_0^t e^{-\alpha\omega u} \sin \omega_d u \, du = \frac{[e^{-\alpha\omega u}(-\alpha\omega \sin \omega_d u - \omega_d \cos \omega_d u)]_0^t}{\alpha^2\omega^2 + \omega_d^2}$$

$$x(t) = \frac{Fe^{-\alpha\omega t}}{m\omega\sqrt{1-\alpha^2}} \cdot \frac{[-\alpha\omega \sin \omega\sqrt{1-\alpha^2}\, t - \omega\sqrt{1-\alpha^2} \cos \omega\sqrt{1-\alpha^2}\, t] + \omega\sqrt{1-\alpha^2}}{\omega^2}$$

As $\omega^2 = k/m$, the corresponding result of exercise 10 is obtained.

Exercise 12: *The system shown in Figure 16 represents an elementary model of a rotating machine of mass $(M - m)$ which is attached to a foundation by a spring k and damper c. The machine has a rotating unbalance of mass m*

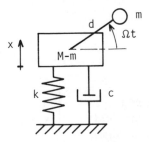

Figure 16

which is located at radial distance d from an axis of rotation about which the radius has a constant angular velocity Ω. Assume that only motion in the x direction is possible. Determine the displacement amplitude X and the amplitude of the force transmitted to the foundation F_t.

The answers are

equation of motion:

$$Mx^{\infty} + cx^{\circ} + kx = md\Omega^2 \sin \Omega t$$

amplitude of displacement:

$$X = \frac{md\Omega^2}{\sqrt{(k - M\Omega^2)^2 + (c\Omega)^2}}$$

amplitude of transmitted force:

$$F_t = md\Omega^2 \sqrt{\frac{k^2 + c^2\Omega^2}{(k - M\Omega^2)^2 + c^2\Omega^2}}$$

Exercise 13: *Develop the force shown in Figure 17 in a Fourier series. The force is applied to the mass of a spring–mass–damper system. Calculate the steady-state motion x(t).*

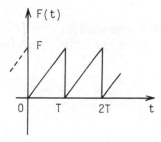

Figure 17

The force expansion is

$$F(t) = F\left(\frac{1}{2} - \frac{1}{\pi} \sum_{p=1}^{\infty} \frac{\sin p\Omega t}{p}\right)$$

and

$$x(t) = \frac{F}{2k} + \sum_{p=1}^{\infty} X_p \sin(p\Omega t - \phi_p)$$

with

$$X_p = \frac{-F/p\pi}{\sqrt{[k - m(p\Omega)^2]^2 + (cp\Omega)^2}}$$

$$\tan \phi_p = \frac{2\alpha(p\Omega/\omega)}{1 - (p\Omega/\omega)^2} \qquad 0 < \phi_p < \pi$$

Exercise 14: *Figure 18 shows a spring–mass system with structural damping. Show that the conditions of amplitude resonance and phase resonance are identical.*

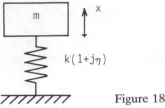

Figure 18

The phase angle equal to $\pi/2$ defines the phase resonance, see equation (106), and it occurs for $\Omega = \omega$.

The amplitude resonance occurs for maximum amplitude, see equation (105), and that can also be shown to occur for $\Omega = \omega$ at which the displacement amplitude is $X_r = F/\eta k$.

Exercise 15: *Calculate the half-power frequency bandwidth for the system in the previous exercise.*

The limits of the bandwidth correspond to one-half of the energy dissipated at resonance. Using equations (88), (105), and the result obtained in exercise 14:

$$\frac{1}{\eta\sqrt{2}} = \frac{1}{\sqrt{[1-(\Omega/\omega)^2]^2 + \eta^2}}$$

from which

$$\left(\frac{\Omega_2}{\omega}\right)^2 = 1 + \eta$$

$$\left(\frac{\Omega_1}{\omega}\right)^2 = 1 - \eta$$

Hence if $\eta \ll 1$,

$$\eta \approx \frac{\Omega_2 - \Omega_1}{\omega} = \frac{\Delta\Omega}{\omega} = \frac{1}{Q}$$

Exercise 16: *The Nyquist diagram. For the system of exercise 14 draw the locus of values of $Z/F = u + jv$ in the complex plane. Show that this locus is a circle passing through the origin and having its center on the imaginary axis. Explain how to obtain ω, η, k, and m from this locus.*

Let

$$\frac{Z}{F} = \frac{1}{k - m\Omega^2 + j\eta k} = u + jv$$

and identify

$$u = \frac{1}{k} \frac{1-(\Omega/\omega)^2}{[1-(\Omega/\omega)^2]^2+\eta^2}$$

$$v = \frac{1}{k} \frac{-\eta}{[1-(\Omega/\omega)^2]^2+\eta^2}$$

These two relations yield the equation of a circle:

$$u^2 + \left(v+\frac{1}{2\eta k}\right)^2 = \frac{1}{4\eta^2 k^2}$$

where Ω increases in a clockwise sense (see Figure 19).

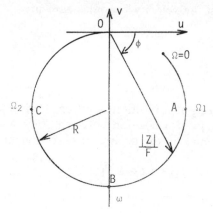

Figure 19

Point B corresponds to the maximum amplitude and therefore at point B, $\Omega=\omega$. Since points A and C are such that $OA=OC=OB/\sqrt{2}$, at point A, $\Omega=\Omega_1$ and at point B, $\Omega=\Omega_2$. Then if, as in exercise 15, η is small,

$$\eta \simeq \frac{\Omega_2-\Omega_1}{\omega}$$

Finally, if R is the radius of the circle:

$$k = \frac{1}{2\eta R}$$

and it follows that the mass of the system is

$$m = \frac{k}{\omega^2}$$

Exercise 17: *Coulomb friction damping. A system consists of a spring and mass supported on a horizontal plane. Let F be the friction force opposing the system motion. At the initial time $t=0$ the system starts from rest with $x_0 = aF/k > 0$. Find the motion of the system and then plot $x(t)$ for $a = 15.5$.*

As the initial position is positive, the equation of motion has the form

$$mx^{\infty} + kx = F$$

with the solution

$$x = \frac{F}{k} + A \sin \omega t + B \cos \omega t$$

The velocity is

$$x^{\circ} = -(a-1)\frac{F}{k} \omega \sin \omega t$$

For this solution to be valid, the velocity must be negative; that is, $a > 1$. This motion will exist until the velocity becomes zero at:

$$\omega t = \pi$$

and at this instant,

$$x = \frac{F}{k}(2-a)$$

If $1 < a < 2$, the mass stops at a position between 0 and F/k.
If $2 < a < 3$, the mass stops at a position between $-F/k$ and 0.
If $a > 3$ the motion continues in the positive x-direction and is governed by the equation

$$mx^{\infty} + kx = -F$$

which can be solved as before.

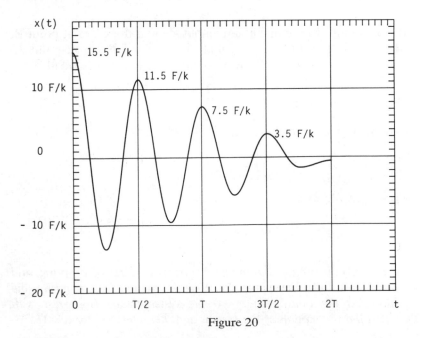

Figure 20

The system is piecewise linear with pieces of length π/ω. The motion of the mass is shown in Figure 20 for $a = 15.5$.

It can be shown that the envelope curves of the maximum values of response are straight lines.

Exercise 18: *Transfer matrices. The transfer matrix links force and displacement at two different points of a system. Determine the transfer matrix of a spring and a mass. Then use these matrices to obtain the frequency of free vibration of a spring–mass system.*

For a system undergoing undamped free vibration, the forces and displacements will be sinusoidal. If F and X are the force and displacement

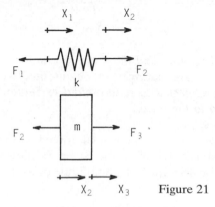

Figure 21

amplitudes at a point in the system (see Figure 21), one has for the spring:

$$F_2 - F_1 = 0$$
$$F_2 = k(X_2 - X_1)$$

for the mass:

$$F_3 - F_2 = -m\omega^2 X_2$$
$$X_3 = X_2$$

The transfer matrices are obtained from these equations and have the following form
  for the spring:

$$\begin{bmatrix} F_2 \\ X_2 \end{bmatrix} = \begin{bmatrix} 1 & 0 \\ \dfrac{1}{k} & 1 \end{bmatrix} \begin{bmatrix} F_1 \\ X_1 \end{bmatrix} = T_k \begin{bmatrix} F_1 \\ X_1 \end{bmatrix}$$

  for the mass:

$$\begin{bmatrix} F_3 \\ X_3 \end{bmatrix} = \begin{bmatrix} 1 & -m\omega^2 \\ 0 & 1 \end{bmatrix} \begin{bmatrix} F_2 \\ X_2 \end{bmatrix} = T_m \begin{bmatrix} F_2 \\ X_2 \end{bmatrix}$$

hence for the system, spring and mass in series,

$$\begin{bmatrix} F_3 \\ X_3 \end{bmatrix} = T_m \cdot T_k \begin{bmatrix} F_1 \\ X_1 \end{bmatrix} = \begin{bmatrix} 1 - \dfrac{m\omega^2}{k} & -m\omega^2 \\ \dfrac{1}{k} & 1 \end{bmatrix} \begin{bmatrix} F_1 \\ X_1 \end{bmatrix}$$

If the spring–mass system is attached at point 1 and free at point 3.

$$F_3 = 0 = \left(1 - \frac{m\omega^2}{k}\right) F_1$$

and since $F_1 \neq 0$, the frequency is

$$\omega = \sqrt{\frac{k}{m}}$$

Exercise 19: *Find the spring and mass transfer matrices for sinusoidal forced vibration. Then use these matrices to obtain the forced response of the spring–mass system subjected to a sinusoidal excitation of amplitude F and frequency Ω applied to the mass.*

It has been shown that in sinusoidal forced vibration, the force and displacement are sinusoidal with the same frequency. Using the same notation as in exercise 18 with F the amplitude of force excitation, one has
for the mass:

$$F_3 - F_2 + F = -m\Omega^2 X_2$$
$$X_3 = X_2$$

and the transfer matrix can be written as:

$$\begin{bmatrix} F_3 \\ X_3 \\ 1 \end{bmatrix} = \begin{bmatrix} 1 & -m\Omega^2 & -F \\ 0 & 1 & 0 \\ 0 & 0 & 1 \end{bmatrix} \begin{bmatrix} F_2 \\ X_2 \\ 1 \end{bmatrix} = T_m^* \begin{bmatrix} F_2 \\ X_2 \\ 1 \end{bmatrix}$$

for the spring:

$$\begin{bmatrix} F_2 \\ X_2 \\ 1 \end{bmatrix} = \begin{bmatrix} 1 & 0 & 0 \\ \dfrac{1}{k} & 1 & 0 \\ 0 & 0 & 1 \end{bmatrix} \begin{bmatrix} F_1 \\ X_1 \\ 1 \end{bmatrix} = T_k^* \begin{bmatrix} F_1 \\ X_1 \\ 1 \end{bmatrix}$$

then

$$\begin{bmatrix} F_3 \\ X_3 \\ 1 \end{bmatrix} = T_m^* \cdot T_k^* \begin{bmatrix} F_1 \\ X_1 \\ 1 \end{bmatrix}$$

and

$$\begin{bmatrix} F_3 \\ X_3 \\ 1 \end{bmatrix} = \begin{bmatrix} 1 - \dfrac{m\Omega^2}{k} & -m\Omega^2 & -F \\ \dfrac{1}{k} & 1 & 0 \\ 0 & 0 & 1 \end{bmatrix} \begin{bmatrix} F_1 \\ X_1 \\ 1 \end{bmatrix}$$

Because of the boundary conditions

$$F_3 = 0 = \left(1 - \frac{m\Omega^2}{k}\right) F_1 - F$$

and

$$X_3 = \frac{F_1}{k}$$

Finally,

$$X_3 = \frac{F}{k(1 - m\Omega^2/k)}$$

and

$$x_3(t) = \frac{F \sin \Omega t}{(k - m\Omega^2)}$$

Exercise 20: *Damped transfer matrices for sinusoidal excitation. Determine the transfer matrices for the parallel spring–viscous damper system shown in Figure 22 by separating the real and imaginary parts of the forces and displacements. Then, for this system attached to a mass m, derive the displacement amplitude of the steady-state motion.*

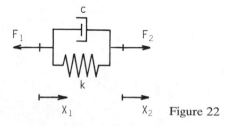

Figure 22

For the parallel spring–viscous damper system shown:

$$F_2 e^{j\Omega t} = k(X_2 e^{j\Omega t} - X_1 e^{j\Omega t}) + jc\Omega(X_2 e^{j\Omega t} - X_1 e^{j\Omega t})$$

Then

$$F_2 = (k + jc\Omega)(X_2 - X_1)$$

Since $F_2 = F_1$,

$$\begin{bmatrix} F_2 \\ X_2 \\ 1 \end{bmatrix} = \begin{bmatrix} 1 & 0 & 0 \\ \dfrac{1}{k+jc\Omega} & 1 & 0 \\ 0 & 0 & 1 \end{bmatrix} \begin{bmatrix} F_1 \\ X_1 \\ 1 \end{bmatrix}$$

and separating the real and imaginary parts,

$$\begin{bmatrix} F_{2r} \\ X_{2r} \\ F_{2i} \\ X_{2i} \\ 1 \end{bmatrix} = \begin{bmatrix} 1 & 0 & 0 & 0 & 0 \\ \dfrac{k}{k^2+c^2\Omega^2} & 1 & \dfrac{c\Omega}{k^2+c^2\Omega^2} & 0 & 0 \\ 0 & 0 & 1 & 0 & 0 \\ \dfrac{-c\Omega}{k^2+c^2\Omega^2} & 0 & \dfrac{k}{k^2+c^2\Omega^2} & 1 & 0 \\ 0 & 0 & 0 & 0 & 1 \end{bmatrix} \begin{bmatrix} F_{1r} \\ X_{1r} \\ F_{1i} \\ X_{1i} \\ 1 \end{bmatrix}$$

The transfer matrix for the mass is found in a similar manner, and is equal to

$$\begin{bmatrix} F_{3r} \\ X_{3r} \\ F_{3i} \\ X_{3i} \\ 1 \end{bmatrix} = \begin{bmatrix} 1 & -m\Omega^2 & 0 & 0 & -F_r \\ 0 & 1 & 0 & 0 & 0 \\ 0 & 0 & 1 & -m\Omega^2 & -F_i \\ 0 & 0 & 0 & 1 & 0 \\ 0 & 0 & 0 & 0 & 1 \end{bmatrix} \begin{bmatrix} F_{2r} \\ X_{2r} \\ F_{2i} \\ X_{2i} \\ 1 \end{bmatrix}$$

Because of the boundary conditions, F_{1r} and F_{1i} are obtained as a function of the components of the excitation force F_r and F_i. Then X_{3r} and X_{3i} can be obtained.

2
Two Degree-of-Freedom Systems

Two degree-of-freedom systems, even though included in N degree-of-freedom systems, are treated separately. This is because their small size allows analytical solution, understanding of more general methods, and an introduction to the concept of coupling. In addition, they provide an explanation of useful applications such as the dynamic vibration absorber.

Certain properties of a vibrating sysem used here will not be proven until the next chapter and, as a preparation for the next chapter, the modal method is used even though direct calculations are simpler.

The contents of the chapter are as follows:
2.1 Undamped systems
2.2 Damped systems
2.3 Vibration absorber
2.4 Exercises.

2.1 Undamped Systems

These are illustrated by the example of Figure 1. The two masses rest on a frictionless horizontal plane.

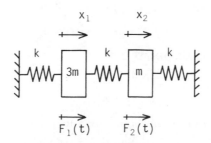

Figure 1 Two degree-of-freedom undamped system

From Newtonian mechanics, the two differential equations of motion are

$$3mx_1^{\circ\circ} + 2kx_1 - kx_2 = F_1(t)$$
$$mx_2^{\circ\circ} + 2kx_2 - kx_1 = F_2(t) \tag{1}$$

where $F_1(t)$ and $F_2(t)$ are the forcing functions acting respectively on masses $3m$ and m.

The equations (1) can also be obtained from Lagrange's equations (see the Appendix). For this purpose, the kinetic energy T, strain energy U, and virtual work of the external forces δW are

$$T = \frac{3mx_1^{\circ 2}}{2} + \frac{mx_2^{\circ 2}}{2} \tag{2}$$

$$U = \frac{kx_1^2}{2} + \frac{k(x_2 - x_1)^2}{2} + \frac{kx_2^2}{2} \tag{3}$$

$$\delta W = F_1(t)\,\delta x_1 + F_2(t)\,\delta x_2 \tag{4}$$

Equations (1) can be written more compactly in the following form:

$$Mx^{\circ\circ} + Kx = F(t) \tag{5}$$

with

$$M = \begin{bmatrix} 3m & 0 \\ 0 & m \end{bmatrix} \quad K = \begin{bmatrix} 2k & -k \\ -k & 2k \end{bmatrix} \tag{6}$$

being the mass matrix and stiffness matrix and

$$x = \begin{bmatrix} x_1 \\ x_2 \end{bmatrix} \quad F(t) = \begin{bmatrix} F_1(t) \\ F_2(t) \end{bmatrix} \tag{7}$$

being the displacement and external force vectors.

2.1.1 Free vibration

In the case of $F(t) = 0$, equations (1) become

$$3mx_1^{\circ\circ} + 2kx_1 - kx_2 = 0$$
$$mx_2^{\circ\circ} + 2kx_2 - kx_1 = 0 \tag{8}$$

Solutions are sought in the form

$$x_1 = X_1 e^{rt}$$
$$x_2 = X_2 e^{rt} \tag{9}$$

Substituting (9) in (8) gives the homogeneous equations

$$(3mr^2 + 2k)X_1 - kX_2 = 0$$
$$(mr^2 + 2k)X_2 - kX_1 = 0 \tag{10}$$

In matrix form, these equations are

$$[r^2M+K]\begin{bmatrix}X_1\\X_2\end{bmatrix}=\Delta(r^2)\begin{bmatrix}X_1\\X_2\end{bmatrix}=0 \tag{11}$$

with

$$\Delta(r^2)=\begin{bmatrix}3mr^2+2k & -k\\-k & mr^2+2k\end{bmatrix} \tag{12}$$

The trivial solution $X_1=X_2=0$ is not of interest. The nontrivial solution is associated with the condition $\Delta(r^2)=0$. The expansion of the determinant gives

$$(3mr^2+2k)(mr^2+2k)-k^2=0 \tag{13}$$

and the two roots are

$$r_1^2=-0.4514\frac{k}{m}$$
$$r_2^2=-2.215\frac{k}{m} \tag{14}$$

From (14),

$$r_1=\pm j0.6719\sqrt{\frac{k}{m}}$$
$$=\pm j\omega_1$$
$$r_2=\pm j1.488\sqrt{\frac{k}{m}} \tag{15}$$
$$=\pm j\omega_2$$

where ω_1 and ω_2 are the frequencies of the system.

Substituting ω_1 and ω_2 into (11) gives

$$\Delta(-\omega_1^2)\begin{bmatrix}X_1\\X_2\end{bmatrix}=\Delta(-\omega_1^2)\phi_1=0$$
$$\Delta(-\omega_2^2)\begin{bmatrix}X_1\\X_2\end{bmatrix}=\Delta(-\omega_2^2)\phi_2=0 \tag{16}$$

where ϕ_1 and ϕ_2 are the mode shapes of the system.

In a broad sense ω_1, ϕ_1 and ω_2, ϕ_2 characterize respectively the two modes of vibration of the system. The word 'mode' is used throughout this text to describe either the mode of vibration ω_i, ϕ_i or the mode shape ϕ_i.

Relationships between X_1 and X_2 can be obtained from each of equations (16)

$$\text{for } \omega_1: \quad X_2=0.6458X_1$$
$$\text{for } \omega_2: \quad X_2=-4.646X_1 \tag{17}$$

It is important to realize that X_1 and X_2 are not independent because (13) implies that the two equations (10) are no longer independent. The two components X_1 and X_2 can therefore only be determined to within a multiplicative constant. The process of choosing this constant is called normalization and there is no unique way to proceed. For systems with only a few degrees of freedom several choices are possible, for example:

(a) Set one of the components equal to unity, then

$$\phi_1 = \begin{bmatrix} X_1 \\ X_2 \end{bmatrix} = \begin{bmatrix} 1 \\ 0.6458 \end{bmatrix} \qquad \phi_2 = \begin{bmatrix} X_1 \\ X_2 \end{bmatrix} = \begin{bmatrix} 1 \\ -4.646 \end{bmatrix} \tag{18}$$

(b) Set the magnitude of the vector equal to unity, then

$$X_1^2 + X_2^2 = 1 \tag{19}$$

and hence

$$\phi_1 = \begin{bmatrix} 0.8401 \\ 0.5425 \end{bmatrix} \qquad \phi_2 = \begin{bmatrix} 0.2104 \\ -0.9776 \end{bmatrix} \tag{20}$$

(c) Set the matrix product $\phi_i^t M \phi_i$ equal to the total mass of the system, then

$$(3X_1^2 + X_2^2)m = 4m \tag{21}$$

and hence

$$\phi_1 = \begin{bmatrix} 1.082 \\ 0.6986 \end{bmatrix} \qquad \phi_2 = \begin{bmatrix} 0.4034 \\ -1.874 \end{bmatrix} \tag{22}$$

From (9) and (18), the motion in free vibration of the system is

$$x_1(t) = \alpha_1 e^{j\omega_1 t} + \beta_1 e^{-j\omega_1 t} + \alpha_2 e^{j\omega_2 t} + \beta_2 e^{-j\omega_2 t}$$
$$x_2(t) = 0.6458(\alpha_1 e^{j\omega_1 t} + \beta_1 e^{-j\omega_1 t}) - 4.646(\alpha_2 e^{j\omega_2 t} + \beta_2 e^{-j\omega_2 t}) \tag{23}$$

Equation (23) can be rewritten in the more convenient form:

$$x_1(t) = a_1 \sin \omega_1 t + b_1 \cos \omega_1 t + a_2 \sin \omega_2 t + b_2 \cos \omega_2 t$$
$$x_2(t) = 0.6458(a_1 \sin \omega_1 t + b_1 \cos \omega_1 t) - 4.646(a_2 \sin \omega_2 t + b_2 \cos \omega_2 t) \tag{24}$$

where the four constants are determined from the initial conditions. If, at the instant $t = 0$, one takes the initial conditions to be

$$x_1(0) = x_0$$
$$x_2(0) = 0$$
$$x_1^\circ(0) = 0$$
$$x_2^\circ(0) = 0 \tag{25}$$

the four constants are solutions to the set of equations

$$b_1 + b_2 = x_0$$
$$0.6458b_1 - 4.646b_2 = 0$$
$$a_1\omega_1 + a_2\omega_2 = 0$$
$$0.6458a_1\omega_1 - 4.646a_2\omega_2 = 0$$

(26)

After completing this solution,

$$x_1(t) = x_0(0.8780 \cos \omega_1 t + 0.1220 \cos \omega_2 t)$$
$$x_2(t) = x_0(0.5669 \cos \omega_1 t - 0.5669 \cos \omega_2 t)$$

(27)

Figure 2 shows $x_2(t)/x_0$ as a function of $t\sqrt{k/m}$. Note that the solution is not sinusoidal.

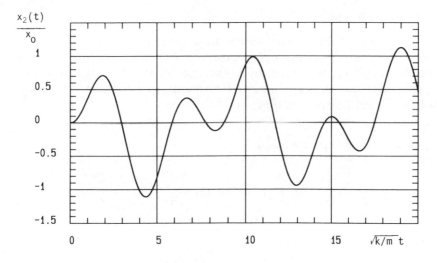

Figure 2   Free motion of mass no. 2 in Figure 1

For the example choosen, it is possible to verify that the modes ϕ_i are related to the mass matrix and the stiffness matrix as follows:

$$\phi_1^t M\phi_2 = 0$$
$$\phi_1^t K\phi_2 = 0$$

(28)

By analogy with the scalar vector product, the ϕ_i are said to be orthogonal with respect to M and K. This orthogonality property will be proven for general systems in the next chapter. In addition, from (16):

$$\omega_1^2 M\phi_1 = K\phi_1$$
$$\omega_2^2 M\phi_2 = K\phi_2$$

(29)

and after premultiplying respectively by ϕ_1^t and ϕ_2^t,

$$\omega_1^2 \phi_1^t M\phi_1 = \phi_1^t K\phi_1$$
$$\omega_2^2 \phi_2^t M\phi_2 = \phi_2^t K\phi_2$$

(30)

The expressions $\phi_1^t M\phi_1$ and $\phi_2^t M\phi_2$ are called the modal masses, and the expressions $\phi_1^t K\phi_1$ and $\phi_2^t K\phi_2$ the modal stiffnesses. If the first method of normalization is used in the above example, one obtains

$$\phi_1^t M\phi_1 = 3.417m$$
$$\phi_2^t M\phi_2 = 24.58m$$
$$\phi_1^t K\phi_1 = 1.542k$$
$$\phi_2^t K\phi_2 = 54.46k$$

(31)

2.1.2 Forced vibration

The transient part of the solution of equation (1) is the general solution of (8). Since engineering systems always possess some damping, this transient solution decays to zero with time. In this section the steady-state solution of (1) is presented. Two cases are discussed: harmonic excitation and excitation which is a general function of time.

Harmonic excitation: direct method

Let

$$F_1(t) = F \sin \Omega t$$
$$F_2(t) = 0$$

(32)

then (1) becomes

$$3mx_1^{\infty} + 2kx_1 - kx_2 = F \sin \Omega t$$
$$mx_2^{\infty} + 2kx_2 - kx_1 = 0$$

(33)

Solutions are sought in the form:

$$x_1(t) = A_1 \sin \Omega t + B_1 \cos \Omega t$$
$$x_2(t) = A_2 \sin \Omega t + B_2 \cos \Omega t$$

(34)

These are then substituted into (33). Since equations (33) must hold for all time, each equation gives two relations corresponding to the vanishing of the coefficients of $\sin \Omega t$ and $\cos \Omega t$. The results are

$$B_1 = B_2 = 0$$

$$A_1 = \frac{F(2k - m\Omega^2)}{(2k - 3m\Omega^2)(2k - m\Omega^2) - k^2} \quad , \quad A_2 = \frac{Fk}{(2k - 3m\Omega^2)(2k - m\Omega^2) - k^2}$$

(35)

hence

$$x_1(t) = \frac{F(2k - m\Omega^2)\sin\Omega t}{(2k - 3m\Omega^2)(2k - m\Omega^2) - k^2} \quad , \quad x_2(t) = \frac{Fk\sin\Omega t}{(2k - 3m\Omega^2)(2k - m\Omega^2) - k^2}$$
(36)

The values of Ω for which the denominators of $x_1(t)$ and $x_2(t)$ vanish correspond to the frequencies ω_1 and ω_2 found for free vibration. If the system is subjected to a harmonic force whose frequency is equal to ω_1 or ω_2, the amplitude of the response will approach infinity. This is the phenomenon of undamped resonance. However, if some damping is included, the constants B_1 and B_2 will be nonzero, and the amplitude of response at resonance will be finite.

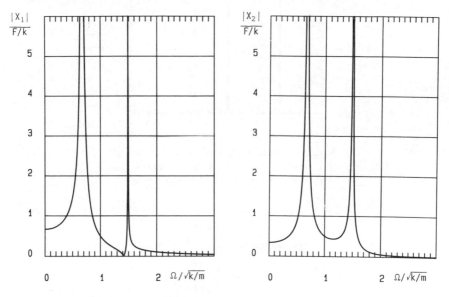

Figure 3   Steady-state amplitude responses of the system of Figure 1 for a sinusoidal forcing function applied to mass no. 1

Figure 3 shows $|X_1|/(F/k)$ and $|X_2|/(F/k)$ as a function of $\Omega/\sqrt{k/m}$ in linear coordinates. Figure 4 shows the phases of $x_1(t)$ and $x_2(t)$ as a function of $\Omega/\sqrt{k/m}$. The phases are $\pi/2$ for $\Omega = \omega_1$ and $\Omega = \omega_2$, but the phase of $x_1(t)$ is also $\pi/2$ at the antiresonance ($|X_1| = 0$).

General function of time: modal method

The main advantage of the modal method for two degree-of-freedom systems is to uncouple the equations of motion. This can be accomplished by

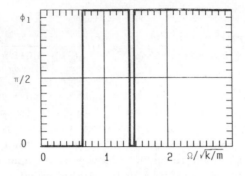

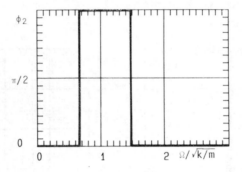

Figure 4 Steady-state phase responses of the system of Figure 1 for a sinusoidal forcing function applied to mass no. 1

using the following change of variable:

$$x = [\phi_1, \phi_2]\begin{bmatrix} q_1 \\ q_2 \end{bmatrix} = \phi q \tag{37}$$

where ϕ_1 and ϕ_2 are the modes of the system and q_1 and q_2 are the new variables. Substituting (37) into (1) and premultiplying by ϕ^t gives

$$\phi^t M \phi q^{\infty} + \phi^t K \phi q = \phi^t F(t) \tag{38}$$

Because of the orthogonality conditions (28), the system (38) consists of two uncoupled differential equations. At this point it is straightforward to obtain $q_1 = q_1(t)$ and $q_2 = q_2(t)$ by proceeding as in section 1.2.3. Finally, $x_1(t)$ and $x_2(t)$ can be recovered by using (37).

The previous example solved by the direct method will now be solved by the modal method. Equation (37) gives

$$\begin{bmatrix} x_1 \\ x_2 \end{bmatrix} = \begin{bmatrix} 1 & 1 \\ 0.6458 & -4.646 \end{bmatrix}\begin{bmatrix} q_1 \\ q_2 \end{bmatrix} \tag{39}$$

and (38) becomes

$$\begin{bmatrix} 3.417m & 0 \\ 0 & 24.58m \end{bmatrix}\begin{bmatrix} q_1^\infty \\ q_2^\infty \end{bmatrix} + \begin{bmatrix} 1.542k & 0 \\ 0 & 54.46k \end{bmatrix}\begin{bmatrix} q_1 \\ q_2 \end{bmatrix} = \begin{bmatrix} F\sin\Omega t \\ F\sin\Omega t \end{bmatrix} \quad (40)$$

The steady-state solutions of the two uncoupled equation in (40) are

$$q_1 = \frac{F\sin\Omega t}{1.542k - 3.417m\Omega^2}, \qquad q_2 = \frac{F\sin\Omega t}{54.46k - 24.58m\Omega^2} \quad (41)$$

and the physical displacements become

$$x_1(t) = q_1 + q_2, \qquad x_2(t) = 0.6458q_1 - 4.646q_2 \quad (42)$$

If Ω is near ω_1 then (42) becomes

$$x_1(t) \simeq q_1$$
$$x_2(t) \simeq 0.6458q_1$$

The system response is then a function of q_1 only. A comparison of x_1 and x_2 shows that only the mode associated with ω_1 appears. A similar conclusion is reached when Ω is near ω_2. This provides a means of measuring the system modes.

For the case of a periodic force represented by a Fourier series, the procedure is the same as above. Let $F_2(t) = 0$ and $F_1(t)$ the force presented in Figure 4 of Chapter 1, then (38) can be written as:

$$3.417mq_1^\infty + 1.542kq_1 = \frac{F_0}{2} + \frac{2F_0}{\pi}\sum_{p=1,3,...}^{\infty}\frac{\sin p\Omega t}{p}$$

$$24.58mq_2^\infty + 54.46kq_2 = \frac{F_0}{2} + \frac{2F_0}{\pi}\sum_{p=1,3,...}^{\infty}\frac{\sin p\Omega t}{p} \quad (43)$$

It is easy to obtain the steady-state solutions, which are

$$q_1(t) = \frac{F_0}{3.086k} + \frac{2F_0}{\pi}\sum_{p=1,3,...}^{\infty}\frac{\sin p\Omega t}{p[1.542k - 3.417m(p\Omega)^2]}$$

$$q_2(t) = \frac{F_0}{108.9k} + \frac{2F_0}{\pi}\sum_{p=1,3,...}^{\infty}\frac{\sin p\Omega t}{p[54.46k - 24.58m(p\Omega)^2]} \quad (44)$$

Then $x_1(t)$ and $x_2(t)$ can be recovered with the aid of (39).

2.2 Damped Systems

Only viscous damping is considered here; a damper of this type is shown in Figure 5. Structural damping is reserved for the exercises. The differential equations of motion of the system can be obtained as in the case of undamped system by using either Newtonian or Lagrangian mechanics. In

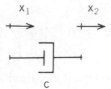

x_1 x_2

Figure 5   Viscous damper

the latter case, the dissipation function R has the form

$$R = \tfrac{1}{2}c(x_2{}^\circ - x_1{}^\circ)^2 \tag{45}$$

The general form of the equation of motion is

$$Mx^\infty + Cx^\circ + Kx = F(t) \tag{46}$$

where C is the viscous damping matrix.

2.2.1   Free vibration

In this case,

$$Mx^\infty + Cx^\circ + Kx = 0 \tag{47}$$

and solutions are sought in the form (9). One finds

$$\Delta(r)\begin{bmatrix} X_1 \\ X_2 \end{bmatrix} = 0 \tag{48}$$

The nontrivial solutions are associated with the condition $\Delta(r) = 0$. The expansion of the determinant gives

$$a_1 r^4 + a_2 r^3 + a_3 r^2 + a_4 r + a_5 = 0 \tag{49}$$

where, in general, the coefficients are nonzero. As these coefficients are real, the roots of (49) are real or in complex conjugate pairs. In the case of small damping, two pairs of complex conjugate roots are obtained:

$$\begin{aligned} r_1 &= -\alpha_1 + j\beta_1 \\ r_2 &= -\alpha_1 - j\beta_1 \\ r_3 &= -\alpha_2 + j\beta_2 \\ r_4 &= -\alpha_2 - j\beta_2 \end{aligned} \tag{50}$$

where α_1 and α_2 are positive. Then $x_1(t)$ may be written as:

$$\begin{aligned} x_1(t) &= e^{-\alpha_1 t}(A_1 \cos \beta_1 t + B_1 \sin \beta_1 t) \\ &\quad + e^{-\alpha_2 t}(A_2 \cos \beta_2 t + B_2 \sin \beta_2 t) \end{aligned} \tag{51}$$

and $x_2(t)$ can be obtained by using the relationship between modes expressed by (48). The quantities α_1 and α_2 characterize the damping and the quantities β_1 and β_2 the frequencies.

In practice, Bairstow's method is often used to find the roots of equations such as (49).

2.2.2 Forced vibration

If the matrix C can be put in the form

$$C = aM + bK \tag{52}$$

where a and b are constants, the damping is said to be proportional. An orthogonality relation identical to (28) can then be shown to hold; that is,

$$\phi_1^t C \phi_2 = a\phi_1^t M \phi_2 + b\phi_1^t K \phi_2 = 0 \tag{53}$$

Hence the modal method can be used as in the case of undamped systems because the differential equations of motion are uncoupled.

In practice, the matrix C is most often nonproportional and therefore $\phi_1^t C \phi_2 \neq 0$.

Harmonic excitation: direct method

The steady-state solutions are sought in the form:

$$\begin{aligned}
x_1(t) &= A_1 \sin \Omega t + B_1 \cos \Omega t \\
x_2(t) &= A_2 \sin \Omega t + B_2 \cos \Omega t
\end{aligned} \tag{54}$$

These are substituted into (46) and the constants A_1, B_1, A_2, B_2, are determined as in the undamped case.

The method has been applied to the system shown in Figure 6, which has the equations of motion:

$$\begin{bmatrix} m & 0 \\ 0 & m \end{bmatrix}\begin{bmatrix} x_1^{\infty} \\ x_2^{\infty} \end{bmatrix} + \begin{bmatrix} 0 & 0 \\ 0 & c \end{bmatrix}\begin{bmatrix} x_1^{\circ} \\ x_2^{\circ} \end{bmatrix} + \begin{bmatrix} 2k & -k \\ -k & 2k \end{bmatrix}\begin{bmatrix} x_1 \\ x_2 \end{bmatrix} = \begin{bmatrix} F \sin \Omega t \\ 0 \end{bmatrix} \tag{55}$$

Plots of $|X_1|/(F/k) = \sqrt{A_1^2 + B_1^2}/(F/k)$ are shown in Figure 7 for three values of the damping coefficient $c = a\sqrt{km}$.

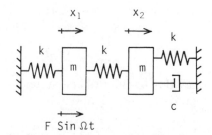

Figure 6 Two degree-of-freedom system with viscous damping

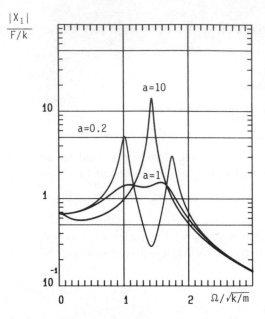

Figure 7   Steady-state amplitude response of
mass no. 1 in Figure 6 for various amounts of
damping

It can be seen from these curves that, for a reasonable value of damping
($a = 0.2$), the response shows two resonant peaks. For a higher value of
damping ($a = 1.0$), the two peaks are barely discernible. Finally, for very
high damping ($a = 10$), mass 2 is nearly motionless and the system is
reduced to that of mass 1 between the two springs of stiffness k which is a
single degree-of-freedom system with small damping. As a result, only one
resonant peak is observed.

2.3   Vibration Absorber

The principle of a vibration absorber is simple and this device is frequently
used to reduce the amplitude of a vibrating system. Let a single degree-of-
freedom system (k_1, m_1) be subjected to a force $F \cos \Omega t$. In steady-state
motion,

$$x_1(t) = \frac{F}{k_1 - m_1 \Omega^2} \cos \Omega t \tag{56}$$

Suppose now that one adds to the original system a second spring–mass
system (k_2, m_2) (see Figure 8). The equations of this combined system are
then

$$m_1 x_1^{\infty} + (k_1 + k_2)x_1 - k_2 x_2 = F \cos \Omega t$$
$$m_2 x_2^{\infty} + k_2 x_2 - k_2 x_1 = 0 \tag{57}$$

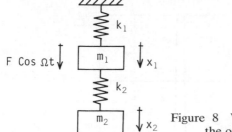

Figure 8 Vibration absorber (k_2, m_2) attached to the original spring–mass system (k_1, m_1)

The steady-state solutions for this undamped system have the form

$$x_1(t) = X_1 \cos \Omega t$$
$$x_2(t) = X_2 \cos \Omega t \tag{58}$$

where

$$X_1 = \frac{F(k_2 - m_2\Omega^2)}{(k_1 + k_2 - m_1\Omega^2)(k_2 - m_2\Omega^2) - k_2^2} \tag{59}$$

$$X_2 = \frac{Fk_2}{(k_1 + k_2 - m_1\Omega^2)(k_2 - m_2\Omega^2) - k_2^2} \tag{60}$$

In particular, for $\sqrt{k_2/m_2}$ chosen equal to Ω, the forcing frequency,

$$X_1 = 0 \quad \text{and} \quad X_2 = \frac{-F}{k_2} \tag{61}$$

Equation (61) shows that for $\sqrt{k_2/m_2} = \Omega$ the motion of the original spring–mass system is completely suppressed. This is the principle of the vibration absorber. In practice, it is more usual to reduce the vibration amplitude of the original system when $\Omega = \sqrt{k_1/m_1}$. Then the vibration of mass m_1 is absorbed if

$$\Omega = \sqrt{\frac{k_1}{m_1}} = \sqrt{\frac{k_2}{m_2}} \tag{62}$$

In using these results for designing vibration absorbers it is necessary to fulfil three requirements:

1. The frequency Ω must be constant or varying only over a small range because the attachment of the absorber splits ω_1 into two resonant frequencies, one on either side of ω_1. Thus, if Ω is too far above or below its design value of ω_1, one will get resonance instead of absorption of its motion.
2. The addition of an auxiliary system to the original system must be technically feasible.
3. The absorber spring (k_2) must be capable of withstanding the force of excitation $F \cos \Omega t$; see equation (61).

50

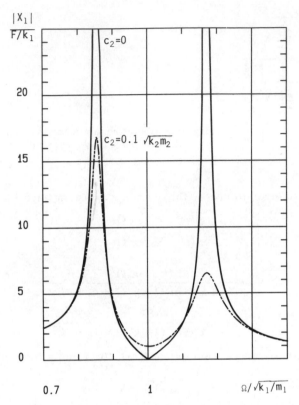

Figure 9 Steady-state amplitude response of the
original spring–mass system (k_1, m_1) in Figure 8 for
two different values of absorber damping (c_2)

Vibration absorbers are sometimes incorporated at the design stage. They
are commonly used to reduce torsional vibration in engines and shafts.

Figure 9 shows the response $|X_1|/(F/k_1)$ as a function of $\Omega/\sqrt{k_1/m_1}$ for the
case where

$$\sqrt{\frac{k_1}{m_1}} = \sqrt{\frac{k_2}{m_2}} \qquad \text{and} \qquad m_1 = 10m_2 \qquad (63)$$

A viscous damper of coefficient $c_2 = 0.1\sqrt{k_2 m_2}$ has also been included to
show a more realistic situation. Notice that damping results in some motion
of the original system (k_1, m_1) at the design conditions.

2.4 Exercises

Exercise 1: *Calculate the frequencies, the associated modes, the modal masses,
and the modal stiffnesses for the system shown in Figure* 10. *Normalize the
modes by setting the first modal component equal to unity.*

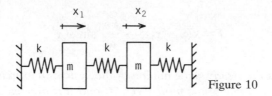

Figure 10

The answers are

$$\omega_1 = \sqrt{\frac{k}{m}}, \qquad \omega_2 = \sqrt{\frac{3k}{m}}$$

$$\phi_1 = \begin{bmatrix} 1 \\ 1 \end{bmatrix}, \qquad \phi_2 = \begin{bmatrix} 1 \\ -1 \end{bmatrix}$$

$$\phi_1^t M \phi_1 = 2m, \qquad \phi_2^t M \phi_2 = 2m$$

$$\phi_1^t K \phi_1 = 2k, \qquad \phi_2^t K \phi_2 = 6k$$

Exercise 2: *Find the free motion $x_1(t)$ and $x_2(t)$ for the system of exercise 1. The initial conditions at $t = 0$ are $x_1 = x_2 = 0$; $x_1^\circ = v_0$; $x_2^\circ = 0$. Also find the force in the spring connecting the two masses.*

The general solutions are

$$x_1(t) = a_1 \sin \omega_1 t + b_1 \cos \omega_1 t + a_2 \sin \omega_2 t + b_2 \cos \omega_2 t$$
$$x_2(t) = a_1 \sin \omega_1 t + b_1 \cos \omega_1 t - a_2 \sin \omega_2 t - b_2 \cos \omega_2 t$$

Applying the initial conditions gives

$$0 = b_1 + b_2$$
$$0 = b_1 - b_2$$
$$v_0 = a_1 \omega_1 + a_2 \omega_2$$
$$0 = a_1 \omega_1 - a_2 \omega_2$$

then

$$b_1 = b_2 = 0$$

$$a_1 \omega_1 = a_2 \omega_2 = \frac{v_0}{2}$$

Hence

$$x_1 = \frac{v_0}{2} \left[\frac{1}{\omega_1} \sin \omega_1 t + \frac{1}{\omega_2} \sin \omega_2 t \right]$$

$$x_2 = \frac{v_0}{2} \left[\frac{1}{\omega_1} \sin \omega_1 t - \frac{1}{\omega_2} \sin \omega_2 t \right]$$

and

$$F = k(x_2 - x_1) = -\frac{kv_0}{\omega_2} \sin \omega_2 t$$

Exercise 3: *Find the steady-state motion of the system in exercise 1. The force of excitation is harmonic of the form F cos Ωt and it is applied to mass 2. Use the direct method.*

The equations are

$$m_1 x_1^{\infty} + 2kx_1 - kx_2 = 0$$
$$mx_2^{\infty} + 2kx_2 - kx_1 = F \cos \Omega t$$

As there is no damping, solutions are sought in the form

$$x_1 = X_1 \cos \Omega t$$
$$x_2 = X_2 \cos \Omega t$$

X_1 and X_2 are solutions of

$$(2k - m\Omega^2)X_1 - kX_2 = 0$$
$$(2k - m\Omega^2)X_2 - kX_1 = F$$

After completing the calculations,

$$x_1 = \frac{Fk \cos \Omega t}{(3k - m\Omega^2)(k - m\Omega^2)}$$

$$x_2 = \frac{F(2k - m\Omega^2) \cos \Omega t}{(3k - m\Omega^2)(k - m\Omega^2)}$$

x_1 and x_2 become infinite for

$$\Omega = \omega_1 = \sqrt{\frac{k}{m}}$$

$$\Omega = \omega_2 = \sqrt{\frac{3k}{m}}$$

Exercise 4: *Repeat exercise 3 but use the modal method.*

The modes, the modal stiffnesses and the modal masses have been calculated in exercise 1. Therefore:

$$\begin{bmatrix} x_1 \\ x_2 \end{bmatrix} = \begin{bmatrix} 1 & 1 \\ 1 & -1 \end{bmatrix} \begin{bmatrix} q_1 \\ q_2 \end{bmatrix}$$

After completing the calculation,

$$2mq_1^{\infty} + 2kq_1 = F \cos \Omega t$$
$$2mq_2^{\infty} + 6kq_2 = -F \cos \Omega t$$

For steady-state motion:

$$q_1 = Q_1 \cos \Omega t$$
$$q_2 = Q_2 \cos \Omega t$$

then

$$q_1 = \frac{F \cos \Omega t}{2k - 2m\Omega^2}$$

$$q_2 = \frac{-F \cos \Omega t}{6k - 2m\Omega^2}$$

and finally

$$x_1 = q_1 + q_2$$
$$x_2 = q_1 - q_2$$

Exercise 5: *Calculate the frequencies, the associated modes, the modal masses and the modal stiffnesses for the system shown in Figure 11. Normalize the mode so that the modulus of each mode is equal to unity.*

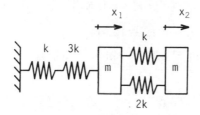

Figure 11

The springs in parallel have an equivalent stiffness of $3k$; those in series have an equivalent stiffness of $3k/4$.

The equations of motion are

$$m x_1^{\infty} + 3.75 k x_1 - 3k x_2 = 0$$
$$m x_2^{\infty} + 3k x_2 - 3k x_1 = 0$$

After completing the calculations, one finds

$$\omega_1 = 0.593 \sqrt{\frac{k}{m}} \qquad \phi_1 = \begin{bmatrix} 0.6618 \\ 0.7497 \end{bmatrix}$$

$$\omega_2 = 2.529 \sqrt{\frac{k}{m}} \qquad \phi_2 = \begin{bmatrix} 0.7497 \\ -0.6618 \end{bmatrix}$$

$$\phi_1^t M \phi_1 = m \qquad \phi_2^t M \phi_2 = m$$
$$\phi_1^t K \phi_1 = 0.3517k \qquad \phi_2^t K \phi_2 = 6.398k$$

Exercise 6: *Rigid-body modes. Calculate the frequencies, the associated modes, the modal stiffnesses, and the modal masses for the system shown in*

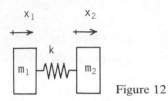

Figure 12

Figure 12. *Normalize the modes by setting the first modal component equal to unity.*

The equations of motion are

$$m_1 x_1{}^{\infty} + k x_1 - k x_2 = 0$$
$$m_2 x_2{}^{\infty} + k x_2 - k x_1 = 0$$

The solutions are found from (9), which gives

$$\begin{bmatrix} m_1 r^2 + k & -k \\ -k & m_2 r^2 + k \end{bmatrix} \begin{bmatrix} X_1 \\ X_2 \end{bmatrix} = 0$$

The expansion of the determinant is as follows:

$$m_1 m_2 r^4 + k(m_1 + m_2) r^2 = 0$$

and the roots are

$$r_1^2 = 0$$

$$r_2^2 = -\frac{k(m_1 + m_2)}{m_1 m_2}$$

Then

$$\omega_1 = 0, \qquad \qquad \phi_1 = \begin{bmatrix} 1 \\ 1 \end{bmatrix}$$

$$\omega_2 = \sqrt{\frac{k(m_1 + m_2)}{m_1 m_2}}, \qquad \phi_2 = \begin{bmatrix} 1 \\ -m_1/m_2 \end{bmatrix}$$

The mode associated with ω_1 is the rigid-body mode. Finally, it can be shown that:

$$\phi_1^t M \phi_1 = m_1 + m_2 \qquad \phi_1^t K \phi_1 = 0$$

$$\phi_2^t M \phi_2 = m_1 + \frac{m_1^2}{m_2} \qquad \phi_2^t K \phi_2 = k\left(1 + \frac{m_1}{m_2}\right)^2$$

Exercise 7: *Consider the same system as in exercise 6 with $m_1 = m_2 = m$. Find $x_1(t)$ and $x_2(t)$ for free vibration using the modal method. The initial conditions at $t = 0$ are $x_1 = x_2 = 0$; $x_1{}^{\circ} = v_0$; $x_2{}^{\circ} = 0$.*

The uncoupled modal equations are

$$2mq_1^{\infty} = 0$$

$$2mq_2^{\infty} + 4kq_2 = 0$$

The solutions for q_1 and q_2 are

$$q_1 = \alpha t + \beta$$

$$q_2 = \gamma \sin \sqrt{\frac{2k}{m}} t + \delta \cos \sqrt{\frac{2k}{m}} t$$

and

$$x_1 = q_1 + q_2$$

$$x_2 = q_1 - q_2$$

Using the initial conditions:

$$0 = \beta + \delta$$

$$0 = \beta - \delta$$

$$v_0 = \alpha + \gamma \sqrt{\frac{2k}{m}}$$

$$0 = \alpha - \gamma \sqrt{\frac{2k}{m}}$$

Then

$$\beta = \delta = 0; \qquad \alpha = \frac{v_0}{2}; \qquad \gamma = \frac{v_0}{2} \sqrt{\frac{m}{2k}}$$

and finally,

$$x_1 = \frac{v_0}{2} \left[t + \sqrt{\frac{m}{2k}} \sin \sqrt{\frac{2k}{m}} t \right]$$

$$x_2 = \frac{v_0}{2} \left[t - \sqrt{\frac{m}{2k}} \sin \sqrt{\frac{2k}{m}} t \right]$$

Exercise 8: *Consider the same system as in exercise 7. Find the steady-state motion $x_1(t)$ and $x_2(t)$ by the direct method. The forcing function is $F \sin \Omega t$ and is applied to mass m_2.*

The equations of motion are

$$mx_1^{\infty} + kx_1 - kx_2 = 0$$

$$mx_2^{\infty} + kx_2 - kx_1 = F \sin \Omega t$$

The solutions have the form

$$x_1 = X_1 \sin \Omega t$$

$$x_2 = X_2 \sin \Omega t$$

Substituting into the equations of motion gives

$$(k - m\Omega^2)X_1 - kX_2 = 0$$
$$(k - m\Omega^2)X_2 - kX_1 = F$$

After completing the necessary calculations,

$$x_1 = \frac{-Fk}{m\Omega^2(2k - m\Omega^2)} \sin \Omega t$$

$$x_2 = \frac{-F(k - m\Omega^2)}{m\Omega^2(2k - m\Omega^2)} \sin \Omega t$$

Note that $x_1(t)$ and $x_2(t)$ become infinite for $\Omega = 0$ and $\Omega = \sqrt{2k/m}$.

Exercise 9: *The behavior of a two degree-of-freedom system is described by the following equations of motion:*

$$\begin{bmatrix} m & 0 \\ 0 & 3m \end{bmatrix}\begin{bmatrix} x_1^{\infty} \\ x_2^{\infty} \end{bmatrix} + \begin{bmatrix} 3k & -k \\ -k & 3k \end{bmatrix}\begin{bmatrix} x_1 \\ x_2 \end{bmatrix} = \begin{bmatrix} 0 \\ F \sin \Omega t \end{bmatrix}$$

where the frequencies and modes can be shown to be

$$\omega_1 = 0.9194 \sqrt{\frac{k}{m}} \quad \phi_1 = \begin{bmatrix} 1 \\ 2.155 \end{bmatrix}, \quad \omega_2 = 1.776 \sqrt{\frac{k}{m}} \quad \phi_2 = \begin{bmatrix} 1 \\ -0.1547 \end{bmatrix}$$

Find the steady-state response if $\Omega \approx \omega_2$.

It is sufficient to write the equation corresponding to q_2:

$$\phi_2^t M\phi_2 q_2^{\infty} + \phi_2^t K\phi_2 q_2 = \phi_2^t F(t)$$
$$1.072mq_2^{\infty} + 3.381kq_2 = -0.1547F \sin \Omega t$$

which implies

$$q_2 = \frac{-0.1547F \sin \Omega t}{3.381k - 1.072m\Omega^2}$$

Then

$$x_1 = \frac{-0.1547F \sin \Omega t}{3.381k - 1.072m\Omega^2}$$

$$x_2 = \frac{0.0239F \sin \Omega t}{3.381k - 1.072m\Omega^2}$$

Exercise 10; *Calculate the frequencies and modes of the system shown in Figure 13 using the method of transfer matrices.*

The transfer matrix between points 6 and 1 is

$$\begin{bmatrix} F_6 \\ X_6 \end{bmatrix} = T_k \cdot T_m \cdot T_k \cdot T_m \cdot T_k \begin{bmatrix} F_1 \\ X_1 \end{bmatrix} = T\begin{bmatrix} F_1 \\ X_1 \end{bmatrix} = \begin{bmatrix} t_{11} & t_{12} \\ t_{21} & t_{22} \end{bmatrix}\begin{bmatrix} F_1 \\ X_1 \end{bmatrix}$$

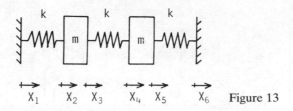

Figure 13

The frequencies are obtained by applying the appropriate boundary conditions. As $X_6 = X_1 = 0$,

$$t_{21}F_1 = 0$$

and ω_1 and ω_2 are the values of ω which make t_{21} vanish. Since

$$t_{21} = \frac{1}{k}\left(1 - \frac{m\omega^2}{k}\right)\left(3 - \frac{m\omega^2}{k}\right)$$

one finds

$$\omega_1 = \sqrt{\frac{k}{m}} \qquad \omega_2 = \sqrt{\frac{3k}{m}}$$

The ratio between X_2 and X_4 establishes the mode:

$$\begin{bmatrix} F_2 \\ X_2 \end{bmatrix} = T_k \begin{bmatrix} F_1 \\ X_1 \end{bmatrix} = \begin{bmatrix} 1 & 0 \\ \frac{1}{k} & 1 \end{bmatrix} \begin{bmatrix} F_1 \\ 0 \end{bmatrix}$$

hence

$$F_2 = F_1 \qquad \text{and} \qquad X_2 = \frac{F_1}{k}$$

Also,

$$\begin{bmatrix} F_4 \\ X_4 \end{bmatrix} = T_{k'} \, T_m \begin{bmatrix} F_2 \\ X_2 \end{bmatrix} = \begin{bmatrix} 1 & -m\omega^2 \\ \frac{1}{k} & 1 - \frac{m\omega^2}{k} \end{bmatrix} \begin{bmatrix} F_2 \\ X_2 \end{bmatrix}$$

and then

$$X_4 = \frac{F_2}{k} + \left(1 - \frac{m\omega^2}{k}\right)X_2 = F_1\left[\frac{1}{k} + \frac{1}{k}\left(1 - \frac{m\omega^2}{k}\right)\right]$$

$$= \frac{F_1}{k}\left(2 - \frac{m\omega^2}{k}\right)$$

For

$$\omega = \omega_1 = \sqrt{\frac{k}{m}}$$

$$X_4 = \frac{F_1}{k} = X_2 \qquad \phi_1 = \begin{bmatrix} 1 \\ 1 \end{bmatrix}$$

and for

$$\omega = \omega_2 = \sqrt{\frac{3k}{m}}$$

$$X_4 = -\frac{F_1}{k} = -X_2 \qquad \phi_2 = \begin{bmatrix} 1 \\ -1 \end{bmatrix}$$

Exercise 11: *Find the steady-state response of the system shown in exercise 10 using the method of transfer matrices. The mass whose motion is $x_2(t) = x_3(t)$ is subjected to a force $F \sin \Omega t$.*

The transfer matrix between points 6 and 1 is

$$\begin{bmatrix} F_6 \\ X_6 \\ 1 \end{bmatrix} = \begin{bmatrix} t_{11}^* & t_{12}^* & t_{13}^* \\ t_{21}^* & t_{22}^* & t_{23}^* \\ t_{31}^* & t_{32}^* & t_{33}^* \end{bmatrix} \begin{bmatrix} F_1 \\ X_1 \\ 1 \end{bmatrix}$$

From the boundary conditions given before, one can obtain F_1:

$$X_6 = t_{21}^* F_1 + t_{23}^* = 0$$

from which

$$F_1 = -\frac{t_{23}^*}{t_{21}^*}$$

Knowing F_1, X_1, one can calculate, successively, $F_2, X_2; F_3, X_3; \ldots$. After all the calculations have been completed,

$$t_{23}^* = -\frac{F}{k} \left(2 - \frac{m\Omega^2}{k} \right)$$

$$t_{21}^* = t_{21} \qquad \text{(see exercise 10 where } \Omega = \omega\text{).}$$

and

$$F_1 = \frac{F(2 - m\Omega^2/k)}{(1 - m\Omega^2/k)(3 - m\Omega^2/k)}$$

$$X_2 = \frac{F_1}{k} = \frac{F}{k} \frac{2 - m\Omega^2/k}{(1 - m\Omega^2/k)(3 - m\Omega^2/k)}$$

$$X_4 = \frac{F}{k} \frac{1}{(1 - m\Omega^2/k)(3 - m\Omega^2/k)}$$

Finally,

$$x_2(t) = X_2 \sin \Omega t$$

$$x_4(t) = X_4 \sin \Omega t$$

Exercise 12: *Find the response for the vibration absorber using complex notation.*

The equations of motion

$$m_1 y_1^{\infty} + (k_1 + k_2) y_1 - k_2 y_2 = F \sin \Omega t$$
$$m_2 y_2^{\infty} + k_2 y_2 - k_2 y_1 = 0$$

can be combined with equations (57) so that

$$m_1 z_1^{\infty} + (k_1 + k_2) z_1 - k_2 z_2 = F e^{j\Omega t}$$
$$m_2 z_2^{\infty} + k_2 z_2 - k_2 z_1 = 0$$

then x_1 and x_2 are the real parts of z_1 and z_2.

The solutions have the form:

$$z_1 = Z_1 e^{j\Omega t}$$
$$z_2 = Z_2 e^{j\Omega t}$$

and by elimination,

$$Z_1 = \frac{F(k_2 - m_2 \Omega^2)}{(k_1 + k_2 - m_1 \Omega^2)(k_2 - m_2 \Omega^2) - k_2^2}$$

which is a real quantity.

Then the motion of the base mass is

$$x_1 = Z_1 \cos \Omega t$$

Exercise 13: *A two degree-of-freedom system $(2m, k, 2m)$ is mounted on a rigid support by a spring of stiffness ak; see Figure 14. The two frequencies of the complete system are ω_1, ω_2, the nonzero frequency of the initial system is ω. Use the computer program no. 5 (see Chapter 7) to calculate ω_1/ω and ω_2/ω for the parameter a in the range $10^{-2} < a < 1$.*

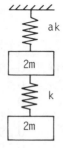

Figure 14

From the results of exercise 6,

$$\omega = \sqrt{\frac{k}{m}}$$

The results from the computer analysis are as tabulated below:

a	1	0.5	0.2	0.1	0.05	0.02	0.01
ω_1/ω	0.44	0.33	0.22	0.16	0.11	0.07	0.05
ω_2/ω	1.14	1.068	1.026	1.013	1.006	1.003	1.001

Note that for $a < 0.1$, $\omega_2 \approx \omega$. Therefore, a vibration experiment can be performed on a free system by suspending it on a weak spring without the support appreciably affecting the results.

Exercise 14: *A single degree-of-freedom system with frequency $\omega_2 = \sqrt{k_2/m_2}$ is mounted on another single degree-of-freedom system having frequency $\omega_1 = \sqrt{k_1/m_1}$, as illustrated in Figure 15. The resulting two degree-of-freedom system has frequencies ω_1^*, ω_2^*. Using computer program no. 5, calculate $(\omega_2^* - \omega_2)/\omega_2$ as a function of ω_2/ω_1 and m_2/m_1.*

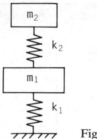

Figure 15

The results of the computer calculation for $(\omega_2^* - \omega_2)/\omega_2$ are as tabulated below:

m_2/m_1 \ ω_2/ω_1	1	1.5	2	3	10	100
0.001	0.016	0.0009	0.0007	0.0006	0.0005	0.0005
0.003	0.029	0.003	0.002	0.0017	0.0015	0.0015
0.01	0.05	0.009	0.007	0.006	0.005	0.005
0.03	0.09	0.025	0.02	0.016	0.015	0.015
0.1	0.17	0.08	0.06	0.05	0.05	0.05
0.3	0.33	0.20	0.17	0.15	0.14	0.14
1	0.62	0.50	0.46	0.43	0.42	0.41

Values of ω_2/ω_1 and m_2/m_1 (or values of k_1, m_1 for k_2, m_2 constant) which are in the upper right-hand corner of this table result in $\omega_2^* \approx \omega_2$. This implies that with judiciously chosen values of k_1 and m_1, it is possible to 'soft' mount a system k_2, m_2 and still measure its 'hard' mounted, or fixed-base, frequency.

Exercise 15: *Torsional vibration absorber. In Figure 16 a hollow disk of inertia I_1 is coupled to a solid disk of inertia I_2 by a liquid. The damping coefficient of the resulting viscous coupling element is c. The torsional stiffness of the elastic shaft is k. The hollow disk is subjected to a harmonic couple of amplitude Γ. Write the equations of motion for the system and find the amplitude of response, Θ_1. Use complex notation. For $I_1/I_2 = 2$, plot $|\Theta_1|/(\Gamma/k)$ as a function of $\Omega/\sqrt{k/I_1}$ for values of viscous damping factor equal to $a = c/\sqrt{kI_1} = 0.05, 0.5$ and 10.*

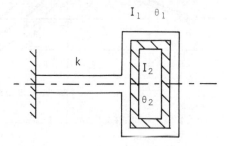

Figure 16

The equations of motion are

$$I_1\theta_1^{\circ\circ} + c\theta_1^{\circ} + k\theta_1 - c\theta_2^{\circ} = \Gamma e^{j\Omega t}$$

$$I_2\theta_2^{\circ\circ} + c\theta_2^{\circ} - c\theta_1^{\circ} = 0$$

Solutions are sought in the form

$$\theta_1 = \Theta_1 e^{j\Omega t}$$

$$\theta_2 = \Theta_2 e^{j\Omega t}$$

from which

$$\Theta_1 = \frac{\Gamma(jc\Omega - I_2\Omega^2)}{I_2\Omega^2(I_1\Omega^2 - k) + jc\Omega(k - I_1\Omega^2 - I_2\Omega^2)}$$

$$= |\Theta_1|e^{-j\phi}$$

and:

$$|\Theta_1| = \Gamma\sqrt{\frac{I_2^2\Omega^4 + c^2\Omega^2}{I_2^2\Omega^4(I_1\Omega^2 - k)^2 + c^2\Omega^2(k - I_1\Omega^2 - I_2\Omega^2)^2}}$$

One can obtain $|\Theta_1|/(\Gamma/k)$ for different values of a, as illustrated in Figure 17.

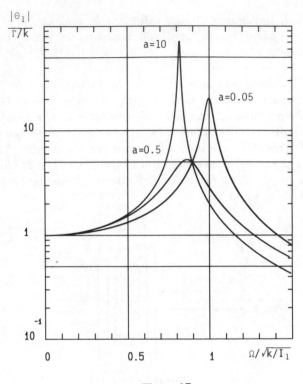

$$\frac{|\Theta_1|}{\Gamma/k}$$

a=10

a=0.05

a=0.5

$\Omega/\sqrt{k/I_1}$

Figure 17

Notice the difference between these response curves and those of Figure 9 in the text. In particular, a viscously coupled absorber does not have a point at which the response of the primary system is brought almost to rest. On the other hand, it does not introduce any new resonances and this allows operation over a wider speed range.

Exercise 16: *Structural damping. Find expressions for the displacement amplitudes of the system shown in Figure 18 by using the direct method with complex notation.*

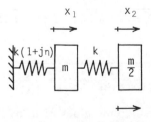

x_1 x_2

$k(1+jn)$ k

m $\frac{m}{2}$

F Cos Ωt   Figure 18

Proceeding as in exercise 12:

$$x_1 = \mathrm{Re}\,[z_1]$$
$$x_2 = \mathrm{Re}\,[z_2]$$

and

$$z_1 = Z_1 e^{j\Omega t}$$
$$z_2 = Z_2 e^{j\Omega t}$$

with

$$F = (F_r + jF_i)e^{j\Omega t} = F_r e^{j\Omega t}$$

Then the equations of motion are

$$-\Omega^2 \begin{bmatrix} m & 0 \\ 0 & \dfrac{m}{2} \end{bmatrix} \begin{bmatrix} Z_1 \\ Z_2 \end{bmatrix} + \begin{bmatrix} 2k & -k \\ -k & k \end{bmatrix} \begin{bmatrix} Z_1 \\ Z_2 \end{bmatrix} + j\eta \begin{bmatrix} k & 0 \\ 0 & 0 \end{bmatrix} \begin{bmatrix} Z_1 \\ Z_2 \end{bmatrix} = \begin{bmatrix} 0 \\ F_r \end{bmatrix}$$

Separating the real parts (Z_{1r}, Z_{2r}) and imaginary parts (Z_{1i}, Z_{2i}) of Z_1 and Z_2 gives

$$\begin{bmatrix} 2k - m\Omega^2 & -k & -\eta k & 0 \\ -k & k - \dfrac{m\Omega^2}{2} & 0 & 0 \\ \eta k & 0 & 2k - m\Omega^2 & -k \\ 0 & 0 & -k & k - \dfrac{m\Omega^2}{2} \end{bmatrix} \begin{bmatrix} Z_{1r} \\ Z_{2r} \\ Z_{1i} \\ Z_{2i} \end{bmatrix} = \begin{bmatrix} 0 \\ F_r \\ 0 \\ 0 \end{bmatrix}$$

For each value of Ω, this system gives $Z_{1r}(\Omega)$, $Z_{2r}(\Omega)$, $Z_{1i}(\Omega)$, and $Z_{2i}(\Omega)$. Then

$$x_1(t) = Z_{1r}(\Omega) \cos \Omega t - Z_{1i}(\Omega) \sin \Omega t$$
$$x_2(t) = Z_{2r}(\Omega) \cos \Omega t - Z_{2i}(\Omega) \sin \Omega t$$

and the response amplitudes are

$$|X_1| = \sqrt{Z_{1r}^2(\Omega) + Z_{1i}^2(\Omega)}$$

$$|X_2| = \sqrt{Z_{2r}^2(\Omega) + Z_{2i}^2(\Omega)}$$

It should be clear that it is a very tedious process to perform these calculations by hand.

Exercise 17: Using program no. 6 of Chapter 7, plot $|X_2|/(F/k)$ as a function of $\Omega/\sqrt{k/m}$ for the results of exercise 16. Consider the cases $\eta = 0.1$, 1.0, and 10.

The results are shown in Figure 19.

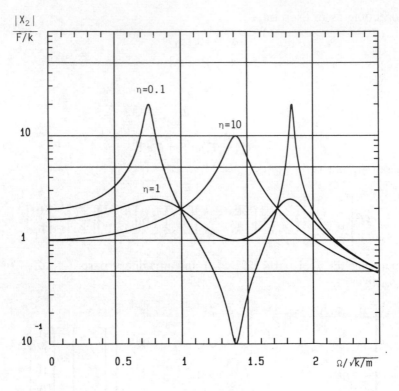

Figure 19

3

N Degree-of-Freedom Systems

The equations of motion for these systems can be obtained from either Newtonian or Lagrangian mechanics. In this chapter, only viscous damping will be considered; however, some of the system properties demonstrated are general and will be used in subsequent chapters. In contrast, some of the numerical methods employed here cannot be used for predicting the vibration behavior of systems having a very large number of degrees of freedom, such as engineering structures which are modeled by the finite element method.

The contents of the chapter are as follows:
3.1 Matrix properties
3.2 Calculation of frequencies and modes
3.3 Response to excitation
3.4 Exercises.

3.1 Matrix Properties

The equations of motion of an N degree-of-freedom system have the form

$$M\ddot{x} + C\dot{x} + Kx = F(t) \tag{1}$$

where M, C, K are, respectively, the mass matrix, the viscous damping matrix, and the stiffness matrix. $F(t)$ and x are the force and displacement vectors.

3.1.1 Symmetry

To simplify the presentation, suppose that each mass of the system has only one degree of freedom.

Mass matrix

The kinetic energy has the form

$$T = \tfrac{1}{2} \sum_{i=1}^{N} m_i \dot{x}_i^{2} \tag{2}$$

and can be written as

$$T = \tfrac{1}{2} x^{ot} M x^{\circ}$$
(3)

with

$$x^{\circ} = (x_1{}^{\circ}, \ldots, x_N{}^{\circ})^t$$

$$M = \begin{bmatrix} m_1 & & \text{zero} \\ & \ddots & \\ \text{zero} & & m_N \end{bmatrix}$$
(4)

where it is noted that the mass matrix M is diagonal. Systems which are modeled by having their mass lumped at discrete points will have diagonal mass matrices. If the linear transformation

$$x = Ay$$
(5)

is substituted into (3), one obtains

$$T = \tfrac{1}{2}(Ay^{\circ})^t M A y^{\circ}$$
$$= \tfrac{1}{2} y^{ot} A^t M A y^{\circ}$$
$$= \tfrac{1}{2} y^{ot} M^* y^{\circ}$$
(6)

where

$$M^* = A^t M A$$
(7)

Symmetry of M^* is guaranteed by the fact that M is premultiplied by A^t and postmultiplied by A.

Stiffness matrix

Let k_{ij} be the stiffness of a spring connecting the x_i to the x_j degree-of-freedom. The corresponding strain energy is

$$U_{ij} = \tfrac{1}{2} k_{ij} (x_i - x_j)^2$$
(8)

From equation (8), one constructs a symmetric submatrix. When this submatrix is inserted into the system stiffness matrix, it appears in a symmetric position. Therefore, the stiffness matrix of the system K is also symmetric. This process of building the matrix of a system from the submatrices of its elements is called assembly. Thus, the system strain energy is given by

$$U = \tfrac{1}{2} x^t K x$$
(9)

If the linear transformation (5) is used, one finds

$$U = \tfrac{1}{2} y^t A^t K A y$$
(10)

and

$$K^* = A^t K A$$
(11)

which is also a symmetric matrix.

Viscous damping matrix

Let c_{ij} be the coefficient of a viscous damper connecting the x_i to the x_j degree of freedom. The dissipation function R_{ij} is

$$R_{ij} = \tfrac{1}{2}c_{ij}(x_i{}^\circ - x_j{}^\circ)^2 \tag{12}$$

As before, it can be shown that the two matrices C and C^* are symmetric, where

$$C^* = A^t C A \tag{13}$$

3.1.2 Positive definite and positive semi-definite matrices

The kinetic energy of the system is always positive unless all the velocities are zero. The mass matrix is then said to be positive definite. The dissipation function R and the strain energy U can be zero with one or more displacements not equal to zero; for example, for rigid-body motion of the system. The matrices C and K are then said to be positive semi-definite.

3.1.3 Properties of the modes

Consider the modes or the mode shapes of the conservative, i.e. undamped, system:

$$Mx^{\infty} + Kx = 0 \tag{14}$$

Orthogonality relations

Solutions to (14) are sought in the form

$$x = Xe^{rt} \tag{15}$$

which gives

$$r^2 M X + K X = 0 \tag{16}$$

or, alternatively,

$$\omega^2 M X = K X \tag{17}$$

in which

$$r = \pm j\omega \tag{18}$$

It can be shown that the solutions ω of (17) are real since the matrix M is symmetric, positive definite and the matrix K is symmetric, positive semi-definite.

Let ω_i, ϕ_i and ω_j, ϕ_j be two solutions of (17); that is,

$$\omega_i^2 M \phi_i = K \phi_i \tag{19}$$

$$\omega_j^2 M \phi_j = K \phi_j \tag{20}$$

Premultiply (19) by ϕ_j^t and (20) by ϕ_i^t. This yields

$$\omega_i^2 \phi_j^t M \phi_i = \phi_j^t K \phi_i \tag{21}$$

$$\omega_j^2 \phi_i^t M \phi_j = \phi_i^t K \phi_j \tag{22}$$

Since M and K are symmetric, equation (22) can be transposed to give

$$\omega_j^2 \phi_j^t M \phi_i = \phi_j^t K \phi_i \tag{23}$$

and combining (21) with (23) gives

$$(\omega_j^2 - \omega_i^2) \phi_j^t M \phi_i = 0 \tag{24}$$

If $\omega_i \neq \omega_j$, which is the most general case, it follows that

$$\phi_j^t M \phi_i = 0 \tag{25}$$

Similarly, from (23), one finds

$$\phi_j^t K \phi_i = 0 \tag{26}$$

Equations (25) and (26) are the orthogonality relations of the modes associated with (14). It follows that if ϕ is the square modal matrix such that

$$\phi = [\phi_1, \ldots, \phi_N] \tag{27}$$

then the matrices $\phi^t M \phi$ and $\phi^t K \phi$ are diagonal.

Consider now the important case of a system having several rigid-body modes for which, of course, the frequencies are zero. For the case of two rigid-body modes, i.e.

$$\omega_1 = 0, \phi_1; \qquad \omega_2 = 0, \phi_2; \qquad \omega_i, \phi_i \qquad i = 3, \ldots, N$$

it follows that:

$$\begin{aligned} K\phi_1 &= 0 \\ K\phi_2 &= 0 \end{aligned} \tag{28}$$

$$\omega_i^2 M \phi_i = K \phi_i \qquad i = 3, \ldots, N \tag{29}$$

As a result of (28):

$$\phi_i^t K \phi_1 = \phi_i^t K \phi_2 = 0 \tag{30}$$

and postmultiplying by ϕ_1 then ϕ_2, the transpose of (29) becomes

$$\phi_i^t M \phi_1 = \phi_i^t M \phi_2 = 0 \tag{31}$$

The classical orthogonality relations among modes, i.e. (25) and (26), are thereby verified between the rigid body modes and the others. In contrast, equation (28) implies that between the two rigid-body modes, one has the relation

$$\phi_1^t K \phi_2 = 0 \tag{32}$$

but, in general

$$\phi_1^t M \phi_2 \neq 0 \qquad (33)$$

Thus rigid-body modes can give rise to off-diagonal terms in the matrix product $\phi^t M \phi$.

Modal mass and modal stiffness

Premultiplying (19) by ϕ_i^t gives

$$\omega_i^2 \phi_i^t M \phi_i = \phi_i^t K \phi_i \qquad (34)$$

or

$$\omega_i^2 = \frac{\phi_i^t K \phi_i}{\phi_i^t M \phi_i} = \frac{k_i}{m_i} \qquad (35)$$

where k_i and m_i are the modal stiffness and modal mass associated with the mode ω_i, ϕ_i.

Viscous damping

The matrix $\phi^t C \phi$ is symmetric. If, in addition, C can be written in the form

$$C = aM + bK \qquad (36)$$

where a and b are constants, the matrix C is said to be proportional. Then $\phi^t C \phi$ is diagonal and

$$\phi_i^t C \phi_i = a \phi_i^t M \phi_i + b \phi_i^t K \phi_i \qquad (37)$$

$$= c_i$$

$$= a m_i + b k_i \qquad (38)$$

and by analogy with the single degree-of-freedom system:

$$\alpha_i = \frac{c_i}{c_{ci}} = \frac{c_i}{2 m_i \omega_i}$$

which can also be written

$$\alpha_i = \frac{a}{2\omega_i} + \frac{b\omega_i}{2} \qquad (39)$$

This property allows an uncoupled solution for system response and permits the definition of the modal damping factors. Notice that this last equation allows any two modal damping factors to be chosen at will, but then all the others are automatically determined.

3.2 Calculation of Frequencies and Modes

Recall from Chapter 1 that small damping has only a second-order effect on frequency. The effect of small damping on modes could also be shown to be small. Therefore, little error is introduced by neglecting damping when it is small.

This section describes methods of finding the frequencies ω_i and modes ϕ_i of an undamped system of N degrees of freedom where the frequencies are arranged so that $\omega_1 < \omega_2 < \ldots < \omega_N$.

3.2.1 Direct method

Solutions are sought in the form

$$x = Xe^{rt} = Xe^{i\omega t} \tag{40}$$

This yields

$$(K - \omega^2 M)X = 0 \tag{41}$$

The procedure is the same as in Chapter 2: the frequencies are obtained by setting the determinant of the matrix $K - \omega^2 M$ equal to zero; then for each ω_i, the components of X are determined and normalized to give ϕ_i. This direct method is efficient only for a few degrees of freedom.

3.2.2 The Rayleigh–Ritz method

This is a generalization by Ritz of Rayleigh's method described in Chapter 1. It is used to reduce the number of degrees of freedom of the system and to estimate the lowest frequency. As before, one begins by making a reasonable hypothesis about the displacement of the system. Ritz suggested that this hypothesis be taken in the form of an expansion; for example, in the form:

$$x = [\gamma_1, \ldots, \gamma_n] \begin{bmatrix} p_1 \\ \vdots \\ p_n \end{bmatrix}$$

$$= \gamma p \tag{42}$$

where γ_i are N-dimensional vectors with $n \ll N$. Equation (42) must satisfy the geometry boundary conditions of the system, i.e. boundary conditions on displacements and slopes. On substituting (42) into (3) and (9),

$$T = \tfrac{1}{2} p^{\circ t} \gamma^t M \gamma p^{\circ}$$
$$U = \tfrac{1}{2} p^t \gamma^t K \gamma p \tag{43}$$

and using Lagrange's equations, one finds

$$\gamma^t M \gamma p^{\circ\circ} + \gamma^t K \gamma p = 0 \tag{44}$$

The order of the system (44) is n, which is much lower than that of the system (1). This reduction in order simplifies the process of finding the frequencies and modes but the solution will not be exact. After solving equations (44) by the direct method, or any other method, the physical displacements are recovered from (42). In fact, it is not obvious how to select the vectors γ_i. A reasonable choice for γ_1 is the static solution of the system subjected to forces equal to the weight of the masses.

In Chapter 5, the Rayleigh–Ritz method will be shown to be very useful in predicting the effects of small changes in structural parameters.

Finally, notice that for $n = 1$ this method reduces to Rayleigh's method.

3.2.3   Iterative method

Iterative methods are very useful. In the iterative method illustrated below, which is only one of many, the frequencies and modes are calculated in succession. Recall equation (17):

$$\omega^2 MX = KX \qquad (45)$$

and rewrite it in the form

$$AX = \frac{1}{\omega^2} X \qquad (46)$$

with

$$A = K^{-1}M$$

Choose a so-called trial vector $\phi_{(1)}$ which is sufficiently general that it can be written as an expansion in terms of all the modes of the system; that is,

$$\phi_{(1)} = a_1 \phi_1 + a_2 \phi_2 + \ldots \qquad (47)$$

with

$$a_1, a_2, \ldots, \neq 0$$

Note the similarity of (47) to an expansion of a n-dimensional cartesian vector in terms of scalar components and unit base vectors. Now the product

$$A\phi_{(1)} = a_1 A\phi_1 + a_2 A\phi_2 + \ldots \qquad (48)$$

when account is taken of (46), that is,

$$A\phi_1 = \frac{\phi_1}{\omega_1^2}; \qquad A\phi_2 = \frac{\phi_2}{\omega_2^2}; \qquad (49)$$

becomes

$$A\phi_{(1)} = \frac{a_1 \phi_1}{\omega_1^2} + \frac{a_2 \phi_2}{\omega_2^2} + \qquad (50)$$

Repeating this process with the new trial vector

$$\phi_{(2)} = A\phi_{(1)} \qquad (51)$$

gives

$$A\phi_{(2)} = a_1 \frac{A\phi_1}{\omega_1^2} + a_2 \frac{A\phi_2}{\omega_2^2} + \ldots$$

$$= a_1 \frac{\phi_1}{\omega_1^4} + a_2 \frac{\phi_2}{\omega_2^4} + \ldots \tag{52}$$

After r iterations, the result is

$$A\phi_{(r)} \simeq \frac{a_1\phi_1}{\omega_1^{2r}} \tag{53}$$

and after $(r+1)$ iterations,

$$A\phi_{(r+1)} \simeq \frac{a_1\phi_1}{\omega_1^{2(r+1)}} \tag{54}$$

Combining (53) and (54) gives

$$A\phi_{(r+1)} \simeq \frac{1}{\omega_1^2} A\phi(r) \tag{55}$$

Equations (54) and (55) allow the determination of ω_1 and ϕ_1. The convergence of this process will be rapid, i.e. r will be small, if $\omega_1 \ll \omega_2$.

The calculation of ω_2, ϕ_2 proceeds in the same manner except that the new trial vector $\phi_{(1)}$ must satisfy

$$\phi_1^t M\phi_{(1)} = 0 \tag{56}$$

Combining (56) and (47) gives

$$\phi_1^t M(a_1\phi_1 + a_2\phi_2 + \ldots) = a_1\phi_1^t M\phi_1 + a_2\phi_1^t M\phi_2 + \ldots = 0 \tag{57}$$

After using the orthogonality relations (25), equation (57) becomes

$$a_1\phi_1^t M\phi_1 = 0 \tag{58}$$

since $\phi_1^t M\phi_1 \neq 0$, it follows that $a_1 = 0$. Therefore condition (56) ensures that $\phi_{(1)}$ contains no component of ϕ_1. The use of this $\phi_{(1)}$ will then lead to a sequence which converges to ω_2 and ϕ_2.

The process of eliminating the component ϕ_1 from the trial vector is called 'sweeping' ϕ_1 out of $\phi_{(1)}$. To determine higher frequencies, it is necessary to use a trial function which has been swept of all previous modes.

The above iterative procedure can also be done using $M^{-1}K$ instead of $K^{-1}M$. The advantage of this is that when M is diagonal, M^{-1} is easy to obtain. The disadvantage is that the procedure will converge to the frequencies in the order $\omega_N, \omega_{N-1}, \ldots, \omega_1$. This alternative is used only for systems having a few degrees of freedom.

The problem of finding the solutions of equations (17) or (46) is an algebraic eigenvalue problem. The eigenvalues and eigenvectors correspond to the frequencies and modes.

3.3 Response to Excitation

Modern methods for calculating the response of N degree-of-freedom systems to excitation when N is large are usually based on a step-by-step method (numerical integration in time) and frequently are preceded by a modal transformation to reduce the number of degrees of freedom. If the motion of interest is the steady-state response to a harmonic force and if N is not too large, the direct method can also be used.

3.3.1 Steady-state response

Direct method for harmonic excitation

For brevity, complex notation is used. Let

$$F(t) = (F_r + jF_i)e^{j\Omega t}$$
$$x(t) = (X_r + jX_i)e^{j\Omega t} \tag{59}$$

Using (59) in (1) gives.

$$-\Omega^2 M(X_r + jX_i) + j\Omega C(X_r + jX_i) + K(X_r + jX_i) = F_r + jF_i \tag{60}$$

Separation of the real and imaginary parts results in

$$\begin{bmatrix} K - \Omega^2 M & -\Omega C \\ \Omega C & K - \Omega^2 M \end{bmatrix} \begin{bmatrix} X_r \\ X_i \end{bmatrix} = \begin{bmatrix} F_r \\ F_i \end{bmatrix} \tag{61}$$

These equations must be solved for each value of Ω and the response constructed using the amplitude and phase of each component of $x(t)$.

This method has disadvantages when N is large; the order of equation (61) is $2N$ and the matrix is no longer symmetric, which causes difficulties with computer calculations. Also, if the damping is small, which is the usual case, it is difficult to obtain the response in the vicinity of resonances because the resonant frequencies are not known.

Modal method

Suppose that the damping is either zero or is proportional. Find the first n frequencies and modes for the undamped system; that is, solve the eigenvalue problem of equation (14) for $\omega_1, \phi_1, \ldots, \omega_n, \phi_n$, with $n \ll N$. This allows the following transformation from physical to modal coordinates:

$$x = [\phi_1, \ldots, \phi_n] \begin{bmatrix} q_1 \\ \vdots \\ q_n \end{bmatrix} \tag{62}$$

$$= \phi q \tag{63}$$

The modal matrix in (62) is a truncated form of the modal matrix defined in (27).

Using (63) in (1) and premultiplying by ϕ^t gives

$$\phi^t M\phi q^\infty + \phi^t C\phi q^\circ + \phi^t K\phi q = \phi^t F(t) \tag{64}$$

According to the orthogonality conditions (25), (26), and (37), equation (64) is a set of n uncoupled differential equations. The first of these equations is

$$m_1 q_1^\infty + c_1 q_1^\circ + k_1 q_1 = F_1(t) \tag{65}$$

with

$$
\begin{aligned}
m_1 &= \phi_1^t M\phi_1 & k_1 &= \phi_1^t K\phi_1 \\
c_1 &= \phi_1^t C\phi_1 = am_1 + bk_1 & F_1(t) &= \phi_1^t F(t)
\end{aligned} \tag{66}
$$

The solution of equations of the type (65) is now straightforward because we can use the results of Chapter 1. Equation (62) is used to recover the physical degrees of freedom. This ability to uncouple the equations of motion is the value of the modal method, and it is a value in both theory and practice.

In many cases of practical interest, the damping is small but is not proportional. In this case, it is possible to use a 'pseudo-modal' method. The coupled modal equations of motion for harmonic excitation have the form:

$$
\begin{bmatrix} m_1 & & \text{zero} \\ & \ddots & \\ \text{zero} & & m_n \end{bmatrix}
\begin{bmatrix} q_1^\infty \\ \vdots \\ \dot{q}_n^\infty \end{bmatrix}
+ \phi^t C\phi
\begin{bmatrix} q_1^\circ \\ \vdots \\ \dot{q}_n^\circ \end{bmatrix}
+
\begin{bmatrix} k & & \text{zero} \\ & \ddots & \\ \text{zero} & & k_n \end{bmatrix}
\begin{bmatrix} q_1 \\ \vdots \\ \dot{q}_n \end{bmatrix}
=
\begin{bmatrix} F_1(t) \\ \vdots \\ F_n(t) \end{bmatrix}
\tag{67}
$$

and they can be solved using the direct method described above because the disadvantages of the direct method identified previously do not exist in this case; that is, $n \ll N$ and the frequencies $\omega_1, \ldots, \omega_n$, at which resonance will occur, have already been found. Of course, the response obtained is valid only for $\Omega < \omega_n$.

For the case of structural damping and harmonic excitation, equation (67) assumes the form

$$-\Omega^2 \phi^t M\phi(Q_r + jQ_i) + \phi^t(K_1 + jK_2)\phi(Q_r + jQ_i) = \phi^t(F_r + jF_i) \tag{68}$$

If the damping is constant throughout the structure, the orthogonality relations will uncouple (68). If this is not the case, the pseudo-modal method is used.

3.3.2 General response

The general response can be obtained by a step-by-step method which gives the response of a system at time t in terms of its response at times prior to t. The method given here is not the most convenient for computation nor does it always converge but it has the advantage of being easy to understand. The step-by-step computer program given in Chapter 7 (program no. 8) uses a related method which can converge.

Develop $x(t)$ in a Taylor series and keep only terms up to the third order, then:

$$x(t) = x(t-\Delta t) + \frac{\Delta t}{1!} x°(t-\Delta t) + \frac{\Delta t^2}{2!} x^{\infty}(t-\Delta t) + \frac{\Delta t^3}{3!} x^{\infty\infty}(t-\Delta t) \quad (69)$$

Similarly, for $x°(t)$ and $x^{\infty}(t)$:

$$x°(t) = x°(t-\Delta t) + \frac{\Delta t}{1!} x^{\infty}(t-\Delta t) + \frac{\Delta t^2}{2!} x^{\infty\infty}(t-\Delta t) \quad (70)$$

$$x^{\infty}(t) = x^{\infty}(t-\Delta t) + \frac{\Delta t}{1!} x^{\infty\infty}(t-\Delta t) \quad (71)$$

This results in an acceleration which is a linear function of the time step Δt. From (69),

$$x^{\infty\infty}(t-\Delta t) = \frac{6}{\Delta t^3}\left[x(t) - x(t-\Delta t) - \Delta t x°(t-\Delta t) - \frac{\Delta t^2}{2} x^{\infty}(t-\Delta t) \right] \quad (72)$$

Substituting (72) in (70) and (71) gives

$$x°(t) = \frac{3}{\Delta t} x(t) - \frac{3}{\Delta t} x(t-\Delta t) - 2x°(t-\Delta t) - \frac{\Delta t}{2} x^{\infty}(t-\Delta t)$$

$$= \frac{3x(t)}{\Delta t} - A_1(t) \quad (73)$$

$$x^{\infty}(t) = \frac{6}{\Delta t^2} x(t) - \frac{6}{\Delta t^2} x(t-\Delta t) - \frac{6}{\Delta t} x°(t-\Delta t) - 2x^{\infty}(t-\Delta t)$$

$$= \frac{6}{\Delta t^2} x(t) - A_2(t) \quad (74)$$

where

$$A_1(t) = \frac{3}{\Delta t} x(t-\Delta t) + 2x°(t-\Delta t) + \frac{\Delta t}{2} x^{\infty}(t-\Delta t) \quad (75)$$

$$A_2(t) = \frac{6}{\Delta t^2} x(t-\Delta t) + \frac{6}{\Delta t} x°(t-\Delta t) + 2x^{\infty}(t-\Delta t) \quad (76)$$

Putting (73) and (74) in (1) gives

$$\left(\frac{6M}{\Delta t^2} + \frac{3C}{\Delta t} + K\right)x(t) = F(t) + MA_2(t) + CA_1(t) \quad (77)$$

If x, $x°$, x^{∞} are known at $(t-\Delta t)$, then $A_1(t)$ and $A_2(t)$ can be determined from (75) and (76) and $x(t)$ can be determined from (77). Finally, (73) and (74) are used to determine $x°(t)$ and $x^{\infty}(t)$. The process is then repeated for $t = t + \Delta t$.

3.4 Exercises

Exercise 1: *Write the equations of motion of the system shown in Figure 1 and calculate the frequencies and modes by the direct method. Normalize the modes by setting the first modal component equal to unity.*

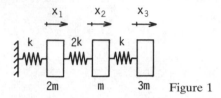

Figure 1

The equations of motion are

$$\begin{bmatrix} 2m & 0 & 0 \\ 0 & m & 0 \\ 0 & 0 & 3m \end{bmatrix}\begin{bmatrix} \overset{\infty}{x_1} \\ \overset{\infty}{x_2} \\ \overset{\infty}{x_3} \end{bmatrix} + \begin{bmatrix} 3k & -2k & 0 \\ -2k & 3k & -k \\ 0 & -k & k \end{bmatrix}\begin{bmatrix} x_1 \\ x_2 \\ x_3 \end{bmatrix} = \begin{bmatrix} 0 \\ 0 \\ 0 \end{bmatrix}$$

If solutions are sought in the form

$$\begin{bmatrix} x_1 \\ x_2 \\ x_3 \end{bmatrix} = \begin{bmatrix} X_1 \\ X_2 \\ X_3 \end{bmatrix} e^{j\omega t}$$

the system of equations becomes

$$\begin{bmatrix} 3k-2m\omega^2 & -2k & 0 \\ -2k & 3k-m\omega^2 & -k \\ 0 & -k & k-3m\omega^2 \end{bmatrix}\begin{bmatrix} X_1 \\ X_2 \\ X_3 \end{bmatrix} = 0$$

The determinant of the square matrix vanishes for

$$\omega_1^2 = 0.1052 \frac{k}{m}$$

$$\omega_2^2 = 0.8086 \frac{k}{m}$$

$$\omega_3^2 = 3.920 \frac{k}{m}$$

For which it can be shown that:

$$\omega_1 = 0.3243 \sqrt{\frac{k}{m}}; \qquad \phi_1 = \begin{bmatrix} 1 \\ 1.395 \\ 2.038 \end{bmatrix}$$

$$\omega_2 = 0.8992 \sqrt{\frac{k}{m}}; \qquad \phi_2 = \begin{bmatrix} 1 \\ 0.6914 \\ -0.4849 \end{bmatrix}$$

$$\omega_3 = 1.980 \sqrt{\frac{k}{m}}; \qquad \phi_3 = \begin{bmatrix} 1 \\ -2.420 \\ 0.2249 \end{bmatrix}$$

Exercise 2: *Consider the same system as in exercise 1. Calculate ω_1 by Rayleigh's method. Assume the displacement pattern obtained by subjecting each mass to a static force proportional to its weight. Normalize the first component of this pattern to unity.*

The deformation pattern is obtained from

$$\begin{bmatrix} 3 & -2 & 0 \\ -2 & 3 & -1 \\ 0 & -1 & 1 \end{bmatrix} \begin{bmatrix} x_1 \\ x_2 \\ x_3 \end{bmatrix} = \begin{bmatrix} 2 \\ 1 \\ 3 \end{bmatrix}$$

and is

$$x_1 = 6; \qquad x_2 = 8; \qquad x_3 = 11$$

Normalizing the x-vector gives

$$\begin{bmatrix} x_1 \\ x_2 \\ x_3 \end{bmatrix} = \begin{bmatrix} 1 \\ 1.333 \\ 1.833 \end{bmatrix} p$$

The expressions for kinetic and strain energy become

$$T = 6.928 m p^{\circ 2}$$
$$U = 0.7359 k p^2$$

and therefore:

$$\omega_1 = 0.3259 \sqrt{\frac{k}{m}}$$

This estimate is high by only 0.5% (see exercise 1).

Exercise 3: *Consider the same problem posed in exercise 2 but now obtain the displacement pattern by subjecting only mass 3 to a static force of unit magnitude.*

This leads to

$$\begin{bmatrix} x_1 \\ x_2 \\ x_3 \end{bmatrix} = \begin{bmatrix} 1 \\ 1.5 \\ 2.5 \end{bmatrix} p$$

$$T = 11.5 m p^{\circ 2}$$
$$U = 1.25 k p^2$$

and

$$\omega_1 = 0.3297 \sqrt{\frac{k}{m}}$$

This estimate is high by 2%.

Exercise 4: *Consider the same system as in exercise 1. Calculate ω_1 and ω_2 by the Rayleigh–Ritz method. The first displacement pattern, γ_1 in equation (42), is the same as in exercise 2. The second pattern, γ_2, is arbitrarily chosen to be $(1, 2, 3)^t$.*

In this case:

$$\begin{bmatrix} x_1 \\ x_2 \\ x_3 \end{bmatrix} = \begin{bmatrix} 1 & 1 \\ 1.333 & 2 \\ 1.833 & 3 \end{bmatrix} \begin{bmatrix} p_1 \\ p_2 \end{bmatrix}$$

and

$$T = \frac{1}{2} \begin{bmatrix} p_1{}^\circ \\ p_2{}^\circ \end{bmatrix}^t \begin{bmatrix} 13.86m & 21.16m \\ 21.16m & 33m \end{bmatrix} \begin{bmatrix} p_1{}^\circ \\ p_2{}^\circ \end{bmatrix}$$

$$U = \frac{1}{2} \begin{bmatrix} p_1 \\ p_2 \end{bmatrix}^t \begin{bmatrix} 1.472k & 2.166k \\ 2.166k & 4k \end{bmatrix} \begin{bmatrix} p_1 \\ p_2 \end{bmatrix}$$

The differential equations of motion become

$$m \begin{bmatrix} 13.86 & 21.16 \\ 21.16 & 33 \end{bmatrix} \begin{bmatrix} p_1{}^\infty \\ p_2{}^\infty \end{bmatrix} + k \begin{bmatrix} 1.472 & 2.166 \\ 2.166 & 4 \end{bmatrix} \begin{bmatrix} p_1 \\ p_2 \end{bmatrix} = 0$$

Solving by the direct method gives the estimates

$$\omega_1 = 0.3249 \sqrt{\frac{k}{m}}$$

$$\omega_2 = 1.098 \sqrt{\frac{k}{m}}$$

Comparison to the exact solutions in exercise 1 shows that the estimate for ω_1 is high by 0.2% and the estimate for ω_2 is high by 20%.

The modes can be calculated as follows.

for ω_1:

$$\begin{bmatrix} p_1 \\ p_2 \end{bmatrix} = \begin{bmatrix} 1 \\ 0.1319 \end{bmatrix}$$

then

$$\begin{bmatrix} X_1 \\ X_2 \\ X_3 \end{bmatrix} = \begin{bmatrix} 1 \\ 1.333 \\ 1.833 \end{bmatrix} + 0.1319 \begin{bmatrix} 1 \\ 2 \\ 3 \end{bmatrix} = \begin{bmatrix} 1.132 \\ 1.597 \\ 2.229 \end{bmatrix}$$

Making the first component unity gives

$$\phi_1 = \begin{bmatrix} 1 \\ 1.411 \\ 1.969 \end{bmatrix}$$

which is very near the exact result.

for ω_2:

$$\begin{bmatrix} p_1 \\ p_2 \end{bmatrix} = \begin{bmatrix} 1 \\ -0.6525 \end{bmatrix}$$

then

$$\begin{bmatrix} X_1 \\ X_2 \\ X_3 \end{bmatrix} = \begin{bmatrix} 1 \\ 1.333 \\ 1.833 \end{bmatrix} - 0.6526 \begin{bmatrix} 1 \\ 2 \\ 3 \end{bmatrix} = \begin{bmatrix} 0.3475 \\ 0.02808 \\ -0.1244 \end{bmatrix}$$

Making the first component unity gives

$$\phi_2 = \begin{bmatrix} 1 \\ 0.0808 \\ -0.3579 \end{bmatrix}$$

This last result is quite different from the exact result. This is to be expected since the value of ω_2 is a distant estimate.

Exercise 5: *Consider the same problem as posed in exercise 4. Use only $\gamma_1 = (1, 1, 1)^t$ and $\gamma_2 = (1, 1.333, 1.833)^t$.*

The answers are

$$\omega_1 = 0.3244 \sqrt{\frac{k}{m}}; \qquad \phi_1 = \begin{bmatrix} 1 \\ 1.412 \\ 2.030 \end{bmatrix}$$

$$\omega_2 = 0.9343 \sqrt{\frac{k}{m}}; \qquad \phi_2 = \begin{bmatrix} 1 \\ 0.4292 \\ -0.4279 \end{bmatrix}$$

Exercise 6: *Find the expressions for kinetic energy and strain energy for the system shown in Figure 2. Find the equations of motion. Calculate the frequencies, modes, modal masses, and modal stiffnesses. Normalize the modes by setting the first modal component equal to unity.*

The energies are

$$2T = mx_1^{o2} + mx_2^{o2} + mx_3^{o2}$$
$$2U = k(x_1 - x_2)^2 + 5k(x_2 - x_3)^2 + k(x_1 - x_3)^2$$

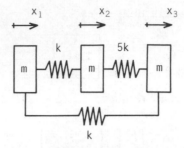

Figure 2

and the equations of motion are

$$\begin{bmatrix} m & 0 & 0 \\ 0 & m & 0 \\ 0 & 0 & m \end{bmatrix}\begin{bmatrix} x_1^{\infty} \\ x_2^{\infty} \\ x_3^{\infty} \end{bmatrix} + \begin{bmatrix} 2k & -k & -k \\ -k & 6k & -5k \\ -k & -5k & 6k \end{bmatrix}\begin{bmatrix} x_1 \\ x_2 \\ x_3 \end{bmatrix} = 0$$

Solutions are sought in the form

$$x = Xe^{j\omega t}$$

and result in

$$m\omega^2(m^2\omega^4 - 14mk\omega^2 + 33k^2) = 0$$

which gives solutions

$$\omega_1 = 0; \qquad \phi_1 = \begin{bmatrix} 1 \\ 1 \\ 1 \end{bmatrix} \qquad \text{with} \qquad \begin{array}{l} \phi_1^t M\phi_1 = 3m \\ \phi_1^t K\phi_1 = 0 \end{array}$$

$$\omega_2 = 1.732\sqrt{\frac{k}{m}}; \qquad \phi_2 = \begin{bmatrix} 1 \\ -0.5 \\ -0.5 \end{bmatrix} \qquad \text{with} \qquad \begin{array}{l} \phi_2^t M\phi_2 = 1.5m \\ \phi_2^t K\phi_2 = 4.5k \end{array}$$

$$\omega_3 = 3.317\sqrt{\frac{k}{m}}; \qquad \phi_3 = \begin{bmatrix} 0 \\ -1 \\ 1 \end{bmatrix} \qquad \text{with} \qquad \begin{array}{l} \phi_3^t M\phi_3 = 2m \\ \phi_3^t K\phi_3 = 22k \end{array}$$

Exercise 7: *Consider the same system as in exercise 1. Calculate, using the results of exercise 1, the matrix products $\phi_i^t K\phi_j$, $\phi_i^t M\phi_j$. Take the first component of the modes equal to unity.*

The results are

$$\phi_1^t K \phi_1 = 1.725k \qquad \phi_1^t M \phi_1 = 16.41m$$
$$\phi_2^t K \phi_2 = 2.574k \qquad \phi_2^t M \phi_2 = 3.183m$$
$$\phi_3^t K \phi_3 = 31.39k \qquad \phi_3^t M \phi_3 = 8.008m$$

All the other products are zero.

Exercise 8: *Consider the same system as in exercise 1. Calculate ω_1 and ϕ_1 by iteration using the trial vector $(1, 1, 1)^t$.*

$$K^{-1} = \frac{1}{k}\begin{bmatrix} 1 & 1 & 1 \\ 1 & \frac{3}{2} & \frac{3}{2} \\ 1 & \frac{3}{2} & \frac{5}{2} \end{bmatrix}$$

$$A = K^{-1}M = \frac{m}{k}\begin{bmatrix} 2 & 1 & 3 \\ 2 & \frac{3}{2} & \frac{9}{2} \\ 2 & \frac{3}{2} & \frac{15}{2} \end{bmatrix} = \frac{m}{k}B$$

Then

$$BX = \frac{k}{m\omega^2}X$$

In what follows, the iteration vector is normalized at each step by making its first component unity.

$$B\phi_{(1)} = \begin{bmatrix} 2 & 1 & 3 \\ 2 & \frac{3}{2} & \frac{9}{2} \\ 2 & \frac{3}{2} & \frac{15}{2} \end{bmatrix}\begin{bmatrix} 1 \\ 1 \\ 1 \end{bmatrix} = \begin{bmatrix} 6 \\ 8 \\ 11 \end{bmatrix} = 6\begin{bmatrix} 1 \\ 1.333 \\ 1.833 \end{bmatrix} = 6\phi_{(2)}$$

$$B\phi_{(2)} = 8.832\begin{bmatrix} 1 \\ 1.385 \\ 2.008 \end{bmatrix} \qquad B\phi_{(3)} = 9.409\begin{bmatrix} 1 \\ 1.393 \\ 2.033 \end{bmatrix}$$

$$B\phi_{(4)} = 9.492\begin{bmatrix} 1 \\ 1.393 \\ 2.036 \end{bmatrix} \qquad B\phi_{(5)} = 9.501\begin{bmatrix} 1 \\ 1.394 \\ 2.036 \end{bmatrix}$$

The mode calculated from $B\phi_{(5)}$ is very near that calculated from $B\phi_{(4)}$ and the number of iterations can therefore be considered sufficient. The frequency is obtained from

$$9.501 = \frac{k}{m\omega_1^2}$$

or

$$\omega_1 = 0.3244 \sqrt{\frac{k}{m}}$$

and the mode is

$$\phi_1 = \begin{bmatrix} 1 \\ 1.394 \\ 2.036 \end{bmatrix}$$

The results for ω_1 and ϕ_1 have been deduced rapidly because, as exercise 1 shows, ω_1 is much smaller than ω_2.

Exercise 9: *Consider the same system as in exercise 1. Calculate ω_2 and ϕ_2 by iteration using the trial vector $(1, 1, 1)^t$.*

The successive vectors $\phi_{(i)}$ are such that

$$[1 \quad 1.394 \quad 2.036] \begin{bmatrix} 2m & 0 & 0 \\ 0 & m & 0 \\ 0 & 0 & 3m \end{bmatrix} \begin{bmatrix} \phi_{(i)1} \\ \phi_{(i)2} \\ \phi_{(i)3} \end{bmatrix} = 0$$

where $\phi_{(i)j}$ is the jth element of $\phi_{(i)}$ at the ith iteration. Expanding the expression gives

$$2\phi_{(i)1} + 1.394\phi_{(i)2} + 6.108\phi_{(i)3} = 0$$

This can be considered in the following form:

$$\begin{bmatrix} \phi_{(i)1}^* \\ \phi_{(i)2}^* \\ \phi_{(i)3}^* \end{bmatrix} = \begin{bmatrix} 0 & -0.6970 & -3.054 \\ 0 & 1 & 0 \\ 0 & 0 & 1 \end{bmatrix} \begin{bmatrix} \phi_{(i)1} \\ \phi_{(i)2} \\ \phi_{(i)3} \end{bmatrix}$$

where $\phi_{(i)}^*$ are now the successive trial vectors. The iteration is then performed using the matrix C:

$$C = B \begin{bmatrix} 0 & -0.6970 & -3.054 \\ 0 & 1 & 0 \\ 0 & 0 & 1 \end{bmatrix} = \begin{bmatrix} 0 & -0.394 & -3.108 \\ 0 & 0.106 & -1.608 \\ 0 & 0.106 & 1.392 \end{bmatrix}$$

Then

$$C\phi_{(1)} = C \begin{bmatrix} 1 \\ 1 \\ 1 \end{bmatrix} = -3.502 \begin{bmatrix} 1 \\ 0.4288 \\ -0.4277 \end{bmatrix}$$

$$C\phi_{(2)} = 1.160 \begin{bmatrix} 1 \\ 0.6319 \\ -0.4740 \end{bmatrix} ; \quad C\phi_{(3)} = 1.224 \begin{bmatrix} 1 \\ 0.6773 \\ -0.4843 \end{bmatrix}$$

$$C\phi_{(4)} = 1.238\begin{bmatrix} 1 \\ 0.6869 \\ -0.4865 \end{bmatrix} ; \qquad C\phi_{(5)} = 1.241\begin{bmatrix} 1 \\ 0.6890 \\ -0.4869 \end{bmatrix}$$

$$C\phi_{(6)} = 1.241\begin{bmatrix} 1 \\ 0.6896 \\ -0.4872 \end{bmatrix}$$

The iteration can now be stopped; the frequency is obtained from

$$1.241 = \frac{k}{m\omega_2^2}$$

and

$$\omega_2 = 0.8977\sqrt{\frac{k}{m}}; \qquad \phi_2 = \begin{bmatrix} 1 \\ 0.6896 \\ -0.4872 \end{bmatrix}$$

Exercise 10: *Structural modification. Consider the same system as in exercise 1 except that mass 2 is modified to have a mass of 1.3m. Using Rayleigh's method, calculate the third frequency of the modified system, ω_3^*.*

This calculation can be done easily because the exact mode corresponding to ω_3 is known from exercise 1. The additional kinetic energy due to the mass addition $0.3m$ is

$$\Delta T = \tfrac{1}{2}0.3m(-2.42)^2 p^{o2} = 0.8785 mp^{o2}$$

Then using the results of exercise 7, the kinetic energy of the modified system is

$$T^* = 4.004 mp^{o2} + 0.8785 mp^{o2}$$
$$= 4.882 mp^{o2}$$

Since the strain energy is constant

$$\omega_3^* = \omega_3 \sqrt{\frac{T}{T^*}} = 1.793\sqrt{\frac{k}{m}}$$

The exact values ω_3^{**} and ϕ_3^{**} of the modified system can be shown to be:

$$\omega_3^{**} = 1.807\sqrt{\frac{k}{m}}; \qquad \phi_3^{**} = \begin{bmatrix} 1 \\ -1.766 \\ 0.2007 \end{bmatrix}$$

so it can be seen that this technique gives convenient estimates of the effect on frequency of small changes of a structure.

Exercise 11: *Consider the same system as in exercise 1. Find the steady-state response to a force $F \sin \Omega t$ acting on mass m. Use the modal method and plot $|X_1|/(F/k)$ as a function of $\Omega/\sqrt{k/m}$.*

The results of exercises 1 and 7 are used to obtain the three uncoupled equations:

$$16.41mq_1^{\infty} + 1.725kq_1 = 1.395F\sin\Omega t$$
$$3.183mq_2^{\infty} + 2.574kq_2 = 0.6914F\sin\Omega t$$
$$8.008mq_3^{\infty} + 31.39kq_3 = -2.420F\sin\Omega t$$

The steady-state solution for q_1 is

$$q_1 = \frac{1.395F\sin\Omega t}{1.725k - 16.41m\Omega^2}$$

or

$$q_1 = \frac{0.8087F\sin\Omega t}{k[1-(\Omega/\omega_1)^2]}$$

Similarly,

$$q_2 = \frac{0.2686F\sin\Omega t}{k[1-(\Omega/\omega_2)^2]}$$

$$q_3 = \frac{-0.07709F\sin\Omega t}{k[1-(\Omega/\omega_3)]^2}$$

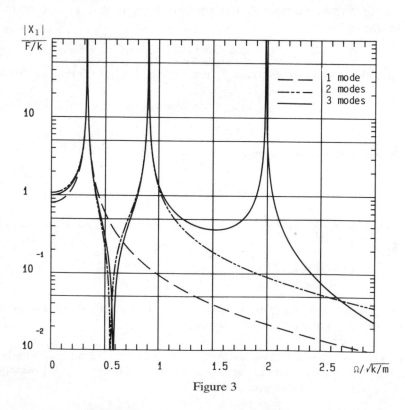

Figure 3

Then

$$x_1 = \left[\frac{0.8087}{1-(\Omega/\omega_1)^2} + \frac{0.2686}{1-(\Omega/\omega_2)^2} - \frac{0.07709}{1-(\Omega/\omega_3)^2}\right]\frac{F\sin\Omega t}{k}$$

$$= X_1 \sin \Omega t$$

Also

$$x_2 = 1.395q_1 + 0.6914q_2 - 2.420q_3$$

The results are plotted in Figure 3.

Exercise 12: *The system shown in Figure 4 is in the vertical plane. The three rods of length L have negligible mass but have a concentrated mass of magnitude m at their tip. The rods are pinned at A, B, and C; their positions are defined by the angular coordinates θ_1, θ_2, and θ_3. The coordinates θ_1, θ_2, θ_3 all remain small during the motion of the system. Take the acceleration of gravity to be g. Write the equations of motion for free vibration. Calculate the frequencies and modes.*

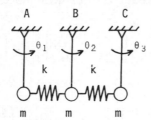

Figure 4

The equations of motion are

$$\begin{bmatrix} mL^2 & 0 & 0 \\ 0 & mL^2 & 0 \\ 0 & 0 & mL^2 \end{bmatrix}\begin{bmatrix} \theta_1^{\infty} \\ \theta_2^{\infty} \\ \theta_3^{\infty} \end{bmatrix}$$

$$+ \begin{bmatrix} mgL+kL^2 & -kL^2 & 0 \\ -kL^2 & mgL+2kL^2 & -kL^2 \\ 0 & -kL^2 & mgL+kL^2 \end{bmatrix}\begin{bmatrix} \theta_1 \\ \theta_2 \\ \theta_3 \end{bmatrix} = 0$$

The frequencies and modes are

$$\omega_1 = \sqrt{\frac{g}{L}}; \quad \phi_1 = \begin{bmatrix} 1 \\ 1 \\ 1 \end{bmatrix} \quad \omega_2 = \sqrt{\frac{g}{L}+\frac{k}{m}}; \quad \phi_2 = \begin{bmatrix} 1 \\ 0 \\ -1 \end{bmatrix}$$

$$\omega_3 = \sqrt{\frac{g}{L}+\frac{3k}{m}}; \quad \phi_3 = \begin{bmatrix} 1 \\ -2 \\ 1 \end{bmatrix}$$

Exercise 13: *Imposed displacement. When a displacement is imposed on a mass point of a spring–mass system, that degree of freedom is lost and the frequencies and modes of the system are changed. Consider the system of exercise 1 with a displacement* $x_3 = X_3 \sin \Omega t$ *imposed on mass 3. Calculate the frequencies and modes and the steady-state response.*

The displacement x_3 is known and the three unknowns are x_1, x_2, and $F(t)$, the force required to generate the displacement x_3 at mass 3. The equations of motion are

$$\begin{bmatrix} 2m & 0 & 0 \\ 0 & m & 0 \\ 0 & 0 & 3m \end{bmatrix} \begin{bmatrix} x_1{}^{\infty} \\ x_2{}^{\infty} \\ -\Omega^2 X_3 \sin \Omega t \end{bmatrix} + \begin{bmatrix} 3k & -2k & 0 \\ -2k & 3k & -k \\ 0 & -k & k \end{bmatrix} \begin{bmatrix} x_1 \\ x_2 \\ X_3 \sin \Omega t \end{bmatrix} = \begin{bmatrix} 0 \\ 0 \\ F(t) \end{bmatrix}$$

Consider just the first two equations:

$$\begin{bmatrix} 2m & 0 \\ 0 & m \end{bmatrix} \begin{bmatrix} x_1{}^{\infty} \\ x_2{}^{\infty} \end{bmatrix} + \begin{bmatrix} 3k & -2k \\ -2k & 3k \end{bmatrix} \begin{bmatrix} x_1 \\ x_2 \end{bmatrix} = \begin{bmatrix} 0 \\ kX_3 \sin \Omega t \end{bmatrix}$$

The frequencies and modes are deduced from the homogeneous form of these equations, i.e.

$$\begin{bmatrix} 2m & 0 \\ 0 & m \end{bmatrix} \begin{bmatrix} x_1{}^{\infty} \\ x_2{}^{\infty} \end{bmatrix} + \begin{bmatrix} 3k & -2k \\ -2k & 3k \end{bmatrix} \begin{bmatrix} x_1 \\ x_2 \end{bmatrix} = \begin{bmatrix} 0 \\ 0 \end{bmatrix}$$

from which

$$\omega_1 = 0.8057 \sqrt{\frac{k}{m}}, \quad \phi_1 = \begin{bmatrix} 1 \\ 0.8508 \end{bmatrix}; \quad \omega_2 = 1.962 \sqrt{\frac{k}{m}}, \quad \phi_2 = \begin{bmatrix} 1 \\ -2.351 \end{bmatrix}$$

The steady-state response can be obtained by seeking solutions of the form

$$x_1 = X_1 \sin \Omega t$$
$$x_2 = X_2 \sin \Omega t$$

One finds

$$\frac{X_1}{X_3} = \frac{2k^2}{(3k - m\Omega^2)(3k - 2m\Omega^2) - 4k^2}$$

$$\frac{X_2}{X_3} = \frac{(3k - 2m\Omega^2)k}{(3k - m\Omega^2)(3k - 2m\Omega^2) - 4k^2}$$

where it is observed that X_1 and X_2 approach infinity for $\Omega = \omega_1$ or $\Omega = \omega_2$.
The force on mass 3 is found from the third equation of motion:

$$F(t) = [-kX_2 + (k - 3m\Omega^2)X_3] \sin \Omega t$$

Some thought will show that if a displacement is imposed on the *base* of a system, the frequencies and modes do not change. For example, if the system of exercise 1 has a displacement $\delta = \Delta \sin \Omega t$ imposed on its base, the equations of motion are the same as for that system with a fixed base and with a *force* $k\Delta \sin \Omega t$ imposed on mass 1.

Exercise 14: *Step-by-step response. The system shown in Figure 5 has a viscous damping coefficient of $c = 0.2\sqrt{km}$. It is subjected to a constant force F at $t = 0$. The initial conditions are $x(0) = 0$, $x°(0) = 0$. Find $x(t)$ in an exact manner and then by the step-by-step method using $\Delta t = T/10$ where T is the period of undamped frequency.*

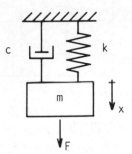

Figure 5

The equation of motion is

$$mx^{\infty} + 0.2\sqrt{km}\, x° + kx = F \qquad (t > 0)$$

Taking account of the initial conditions, the exact solution of this equation is

$$x(t) = \frac{F}{k}[1 - 1.005e^{-0.1\omega t} \sin(0.995\omega t + 1.471)]$$

where the phase angle is in radians.

The expressions utilized in the step-by-step calculation are the following:

$$\frac{6M}{\Delta t^2} + \frac{3C}{\Delta t} + K = 17.15k$$

$$A_1(t) = 4.775\sqrt{\frac{k}{m}}\, x(t - \Delta t) + 2x°(t - \Delta t) + 0.3142\sqrt{\frac{m}{k}}\, x^{\infty}(t - \Delta t)$$

$$A_2(t) = 15.20\frac{k}{m}\, x(t - \Delta t) + 9.549\sqrt{\frac{k}{m}}\, x°(t - \Delta t) + 2x^{\infty}(t - \Delta t)$$

$$x(t) = 0.05830\frac{F}{k} + 0.05830\frac{m}{k}\, A_2(t) + 0.01166\sqrt{\frac{m}{k}}\, A_1(t)$$

$$x°(t) = 4.775\sqrt{\frac{k}{m}}\, x(t) - A_1(t)$$

$$x^{\infty}(t) = 15.20\frac{k}{m}\, x(t) - A_2(t)$$

88

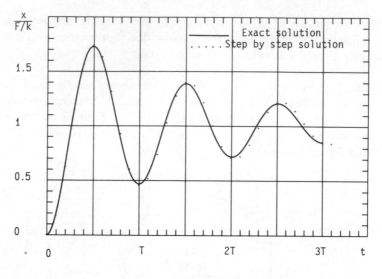

Figure 6

The resultant shown in Figure 6 for step-by-step solution has been calculated with the aid of program no. 8 of Chapter 7 with $\theta = 1$. As the size of the time step Δt is reduced, the step-by-step solution will approach the exact solution.

Exercise 15: *Consider a system of eight degrees of freedom consisting of a chain of identical springs and masses; see Figure 7. It can be shown that the*

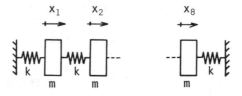

Figure 7

frequencies of this system are $\omega_i = 2\sqrt{k/m}\cos(9-i)\pi/18$, $i = 1, 2, \ldots, 8$.
1. *Calculate the modes ϕ_1 and ϕ_2 corresponding to ω_1 and ω_2.*
2. *The first two springs, beginning from the left, are modified so as to have stiffnesses $1.5k$. Calculate the two lowest frequencies ω_1^* and ω_2^* and their corresponding modes ϕ_1^* and ϕ_2^*. Use program no. 5.*
3. *Recalculate ω_1^*, ω_2^*; ϕ_1^*, ϕ_2^* by the Rayleigh–Ritz method using as displacement patterns the first two modes of the unmodified system. Designate these estimates as ω_1^{**}, ω_2^{**}; ϕ_1^{**}, ϕ_2^{**}.*

For part 1:

$$\omega_1 = 0.3473 \sqrt{\frac{k}{m}};$$

$$\phi_1 = [1, 1.879, 2.532, 2.879, 2.879, 2.532, 1.879, 1]^t$$

$$\omega_2 = 0.6840 \sqrt{\frac{k}{m}};$$

$$\phi_2 = [1, 1.532, 1.347, 0.5321, -0.5321, -1.347, -1.532, -1]^t$$

For part 2:

$$\omega_1^* = 0.3728 \sqrt{\frac{k}{m}};$$

$$\phi_1^* = [1, 1.907, 3.003, 3.682, 3.849, 3.482, 2.630, 1.413]^t$$

$$\omega_2^* = 0.7219 \sqrt{\frac{k}{m}};$$

$$\phi_2^* = [1, 1.653, 1.770, 0.9651, -0.3429, -1.472, -1.834, -1.240]^t$$

For part 3:

For the modified system, one has

$$[\phi_1, \phi_2]^t [M^*] [\phi_1, \phi_2] = m \begin{bmatrix} 38.46 & 0 \\ 0 & 10.89 \end{bmatrix}$$

$$[\phi_1, \phi_2]^t [K^*] [\phi_1, \phi_2] = k \begin{bmatrix} 5.525 & 0.7338 \\ 0.7338 & 5.737 \end{bmatrix}$$

from which ω_1^{**} and ω_2^{**} and the associated modes of this two degree-of-freedom system can be deduced.

$$\omega_1^{**} = 0.3746 \sqrt{\frac{k}{m}} \quad \text{with} \quad \begin{bmatrix} 1 \\ -0.1744 \end{bmatrix};$$

$$\omega_2^{**} = 0.7281 \sqrt{\frac{k}{m}} \quad \text{with} \quad \begin{bmatrix} 1 \\ 20.26 \end{bmatrix}$$

Then the modes of the modified system are:

$$\phi_1^{**} = [1, 1.952, 2.782, 3.375, 3.599, 3.351, 2.599, 1.422]^t$$

and:

$$\phi_2^{**} = [1, 1.548, 1.403, 0.6425, -0.3716, -1.165, -1.372, -0.9060]^t$$

Exercise 16: *A rotating shaft subjected to torsion is modeled by three disks of inertia I, I, I/2 and three rods of identical torsional stiffness k; see Figure 8.*

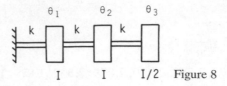

I   I   I/2   Figure 8

The shaft rotates with a constant angular velocity Ω *and* θ_1, θ_2, θ_3 *are measured with respect to the instantaneous shaft position. Write the kinetic and strain energies of this system in terms of* θ_1, θ_2, θ_3. *Calculate the frequencies and modes.*

The kinetic energy associated with shaft rotation is a constant. As such, it makes no contribution to the equations of motion and will, therefore, be ignored in what follows. The kinetic and strain energies are

$$2T = I\dot{\theta_1}^2 + I\dot{\theta_2}^2 + \tfrac{1}{2}I\dot{\theta_3}^2$$
$$2U = k\theta_1^2 + k(\theta_2 - \theta_1)^2 + k(\theta_3 - \theta_2)^2$$

The equations of motion are

$$I\begin{bmatrix} 1 & 0 & 0 \\ 0 & 1 & 0 \\ 0 & 0 & \tfrac{1}{2} \end{bmatrix}\begin{bmatrix} \ddot{\theta_1} \\ \ddot{\theta_2} \\ \ddot{\theta_3} \end{bmatrix} + k\begin{bmatrix} 2 & -1 & 0 \\ -1 & 2 & -1 \\ 0 & -1 & 1 \end{bmatrix}\begin{bmatrix} \theta_1 \\ \theta_2 \\ \theta_3 \end{bmatrix} = 0$$

from which

$$\omega_1 = 0.5176\sqrt{\frac{k}{I}}; \qquad \phi_1 = \begin{bmatrix} 1 \\ 1.732 \\ 2 \end{bmatrix}$$

$$\omega_2 = 1.414\sqrt{\frac{k}{I}}; \qquad \phi_2 = \begin{bmatrix} 1 \\ 0 \\ -1 \end{bmatrix}$$

$$\omega_3 = 1.932\sqrt{\frac{k}{I}}; \qquad \phi_3 = \begin{bmatrix} 1 \\ -1.732 \\ 2 \end{bmatrix}$$

Exercise 17: *Two torsion shafts are connected by a coupling of ratio n,* $\theta_3 = -n\theta_2$; *see Figure 9. The shafts rotate at constant speed. The model is of*

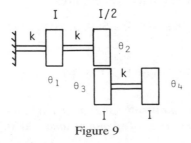

Figure 9

the same kind as used in exercise 16. Write the equations of motion and calculate the frequencies and modes for $n = 2$.

The kinetic and strain energies are

$$2T = I\theta_1^{\circ 2} + \tfrac{1}{2}I\theta_2^{\circ 2} + I\theta_3^{\circ 2} + I\theta_4^{\circ 2}$$
$$= I\theta_1^{\circ 2} + I(\tfrac{1}{2} + n^2)\theta_2^{\circ 2} + I\theta_4^{\circ 2}$$
$$2U = k\theta_1^2 + k(\theta_2 - \theta_1)^2 + k(\theta_4 - \theta_3)^2$$
$$= k\theta_1^2 + k(\theta_2 - \theta_1)^2 + k(\theta_4 + n\theta_2)^2$$

The resulting equations of motion are

$$I\begin{bmatrix} 1 & 0 & 0 \\ 0 & \tfrac{1}{2}+n^2 & 0 \\ 0 & 0 & 1 \end{bmatrix}\begin{bmatrix} \theta_1^{\circ\circ} \\ \theta_2^{\circ\circ} \\ \theta_4^{\circ\circ} \end{bmatrix} + k\begin{bmatrix} 2 & -1 & 0 \\ -1 & 1+n^2 & n \\ 0 & n & 1 \end{bmatrix}\begin{bmatrix} \theta_1 \\ \theta_2 \\ \theta_4 \end{bmatrix} = 0$$

and if $n = 2$:

$$\omega_1 = 0.2358\sqrt{\frac{k}{I}}; \qquad \phi_1 = \begin{bmatrix} 1 \\ 1.944 \\ -4.118 \end{bmatrix}$$

$$\omega_2 = 1.299\sqrt{\frac{k}{I}}; \qquad \phi_2 = \begin{bmatrix} 1 \\ 0.3114 \\ 0.9046 \end{bmatrix}$$

$$\omega_3 = 1.538\sqrt{\frac{k}{I}}; \qquad \phi_3 = \begin{bmatrix} 1 \\ -0.3670 \\ -0.5369 \end{bmatrix}$$

Exercise 18: *The system in exercise 17 is subjected to a couple* $\Gamma \sin \Omega t$ *applied to disk 1. In order to suppress the resonance at* ω_2, *a torsional vibration absorber of inertia* $I/2$ *and stiffness* $0.844k$ *is attached to disk 4; see Figure 10.*

0.844k

θ_4 θ_5

I/2   Figure 10

Plot $|\Theta_4|/(\Gamma/k)$ *with and without the absorber as a function of* $\Omega/\sqrt{k/I}$. *Use program no. 6.*

The response plot is shown in Figure 11.

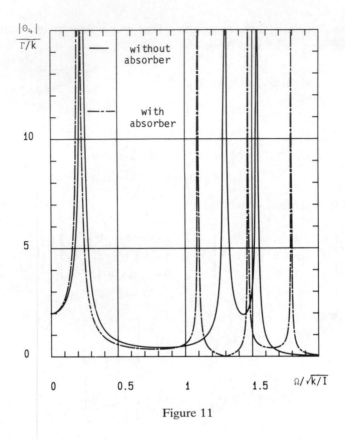

Figure 11

Exercise 19: *Figure* 12 *shows a five degree-of-freedom system which has a chain of identical masses* m *coupled by identical springs* k *and dampers*

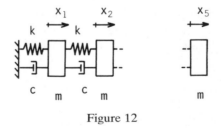

Figure 12

$c = 10^{-2}\sqrt{km}$. *Mass* 3 *is subjected to a forcing function* $F \cos \Omega t$. *Plot* $|X_5|/(F/k)$ *versus* $\Omega/\sqrt{k/m}$. *Use the modal method together with program no.* 7. *First, obtain the plot by using only the first two modes and then by using the first four modes.*

The response plot is shown in Figure 13.

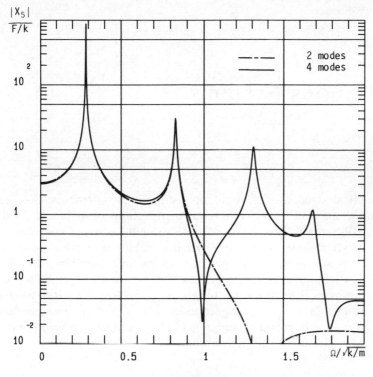

Figure 13

4

Continuous Systems

Many practical structures are built up using simple elements like beams, plates, and shells which have continuous distributions of mass and elasticity. Except possibly for the important case of straight beams of constant cross-section, analytical solutions for the dynamic behavior of these basic structural elements are limited to simple geometries and boundary conditions.

In cases where analytical approaches are laborious, it is better to go directly to numerical methods. In this chapter, the Rayleigh–Ritz method is usually chosen; in the following chapter, the finite element method is utilized. Also in this chapter, damping is not taken into account and only steady-state response is considered.

The contents of the chapter are as follows:
4.1   Equations of motion for bars, rods, and beams
4.2   Frequencies, modes, and orthogonality
4.3   Approximate methods
4.4   Response to excitation
4.5   Kinetic and strain energies of plates
4.6   Frequencies and modes for plates, response to excitation
4.7   Kinetic and strain energies of rotor elements
4.8   Exercises

4.1   Equations of Motion for Bars, Rods, and Beams

In the following sections, the classical equations for straight members with continuous distributions of mass and elasticity are derived for longitudinal, torsional, and bending motion. Continuous systems are the next logical step in this text, since they may be considered as N degree-of-freedom systems with $N \rightarrow \infty$.

4.1.1   Longitudinal motion of a bar

The motion is defined by:
u,   axial displacement
P,   axial force acting on the cross-section

94

P_{ex}, external axial force per unit length
S, cross-sectional area
E, Young's modulus
ρ, mass density.

The application of Newton's laws along the longitudinal axis of the element

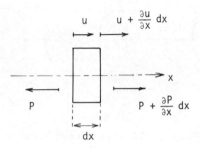

Figure 1 Element of a bar in
longitudinal motion

in Figure 1 gives

$$\rho S \, dx \, \frac{\partial^2 u}{\partial t^2} = P + \frac{\partial P}{\partial x} \, dx - P + P_{ex} \, dx \tag{1}$$

which becomes

$$\rho S \frac{\partial^2 u}{\partial t^2} = \frac{\partial P}{\partial x} + P_{ex} \tag{2}$$

The relation between force and displacement is

$$\frac{P}{S} = E \frac{\partial u}{\partial x} \tag{3}$$

Combining (2) and (3) gives the classical partial differential equation of a bar in longitudinal motion:

$$\rho S \frac{\partial^2 u}{\partial t^2} = \frac{\partial}{\partial x} \left(ES \frac{\partial u}{\partial x} \right) + P_{ex} \tag{4}$$

which for a constant cross-section becomes

$$\rho S \frac{\partial^2 u}{\partial t^2} = ES \frac{\partial^2 u}{\partial x^2} + P_{ex} \tag{5}$$

4.1.2 Torsional motion of a rod

The motion is defined by:
θ,   angle of twist
T,   torsional couple

T_{ex}, external torsional couple per unit length
I_θ, mass moment of inertia about longitudinal axis, per unit length
J, area moment of inertia about the longitudinal axis
G, shear modulus of the rod material.

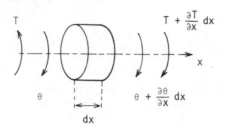

Figure 2   Element of a rod in torsion

The application of Newton's laws about the x direction of the element in Figure 2 gives

$$I_\theta \, dx \frac{\partial^2 \theta}{\partial t^2} = T + \frac{\partial T}{\partial x} \, dx - T + T_{ex} \, dx \tag{6}$$

which becomes

$$I_\theta \frac{\partial^2 \theta}{\partial t^2} = \frac{\partial T}{\partial x} + T_{ex} \tag{7}$$

The relation between torsional couple and the angle of twist is

$$T = GJ \frac{\partial \theta}{\partial x} \tag{8}$$

Combining (7) and (8) gives the classical partial differential equation of a rod in torsion:

$$I_\theta \frac{\partial^2 \theta}{\partial t^2} = \frac{\partial}{\partial x} \left(GJ \frac{\partial \theta}{\partial x} \right) + T_{ex} \tag{9}$$

which for a constant cross-section becomes

$$I_\theta \frac{\partial^2 \theta}{\partial t^2} = GJ \frac{\partial^2 \theta}{\partial x^2} + T_{ex} \tag{10}$$

Note that equations (5) and (10) are mathematically the same. It can be shown that $I_\theta = J\rho$ is valid only for rods of circular cross-section. Using this relationship for other cross-sections, e.g. rectangular, can lead to serious error.

4.1.3 Bending motion of a beam

The motion is defined by:
v, lateral deflection
ψ, slope of neutral axis
Q, lateral shear force
M, bending moment
Q_{ex}, lateral external force per unit length
I, area moment of inertia of beam cross-section about the neutral axis
a, shear factor which appears in the expression for lateral deflection due to shear; it is of the order of unity for beam cross-sections usually found in practice.

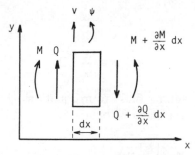

Figure 3 Element of a beam in bending

The application of Newton's laws to the element in Figure 3 in the lateral y-direction and about the z-direction gives

$$\rho S\,dx\,\frac{\partial^2 v}{\partial t^2}=Q-Q-\frac{\partial Q}{\partial x}dx+Q_{ex}\,dx \tag{11}$$

$$\rho I\,dx\,\frac{\partial^2 \psi}{\partial t^2}=-M+M+\frac{\partial M}{\partial x}dx-Q\,dx \tag{12}$$

which simplify to

$$\rho S\frac{\partial^2 v}{\partial t^2}=-\frac{\partial Q}{\partial x}+Q_{ex} \tag{13}$$

$$\rho I\frac{\partial^2 \psi}{\partial t^2}=\frac{\partial M}{\partial x}-Q \tag{14}$$

From the theory of strength of materials, the relations among Q, M, v, ψ are

$$\frac{\partial \psi}{\partial x}=\frac{M}{EI} \tag{15}$$

$$\psi-\frac{Q}{aSG}=\frac{\partial v}{\partial x} \tag{16}$$

The terms $\rho I\, \partial^2\psi/\partial t^2$ and Q/aSG are usually second-order effects. The first is the rotatory inertia introduced by Rayleigh and the second is the shear deformation introduced by Timoshenko. In so-called Timoshenko beams these effects are considered; in Bernoulli–Euler beams they are not.

In the following, we shall ignore these secondary effects and obtain the equations of motion for the Bernoulli–Euler beam. Equations (13), (14), (15), (16) become

$$\rho S\frac{\partial^2 v}{\partial t^2} = -\frac{\partial Q}{\partial x} + Q_{ex}$$

$$\frac{\partial M}{\partial x} - Q = 0$$

$$\frac{\partial \psi}{\partial x} = \frac{M}{EI} \tag{17}$$

$$\psi = \frac{\partial v}{\partial x}$$

Eliminating Q, M, ψ among these four equations gives the classical partial differential equation of a beam in bending:

$$\frac{\partial^2}{\partial x^2}\left(EI\frac{\partial^2 v}{\partial x^2}\right) + \rho S\frac{\partial^2 v}{\partial t^2} - Q_{ex} = 0 \tag{18}$$

which for a constant cross section becomes

$$EI\frac{\partial^4 v}{\partial x^4} + \rho S\frac{\partial^2 v}{\partial t^2} - Q_{ex} = 0 \tag{19}$$

Note that (19) is mathematically different from both (5) and (10).

4.2 Frequencies, Modes, and Orthogonality

Consider the case for which equations (5) and (19) are homogeneous, i.e. $P_{ex} = 0$, $Q_{ex} = 0$. The free-vibration solutions to these equations will be obtained by the method of separation of variables. Since the differential equation of motion for torsion is mathematically identical to that for longitudinal motion, the case of torsion will not be treated.

4.2.1 Longitudinal motion

Let

$$u(x, t) = \phi(x)f(t) \tag{20}$$

and substitute into (5). This gives

$$\rho S\phi(x)\frac{d^2 f(t)}{dt^2} - ESf(t)\frac{d^2\phi(x)}{dx^2} = 0 \tag{21}$$

which can be separated into two ordinary differential equations of motion, one in space and one in time:

$$\frac{E}{\rho}\frac{1}{\phi(x)}\frac{\mathrm{d}^2\phi(x)}{\mathrm{d}x^2} = \frac{1}{f(t)}\frac{\mathrm{d}^2f(t)}{\mathrm{d}t^2} = \text{const.} = -\omega^2 \tag{22}$$

The separation constant has been set equal to $-\omega^2$ so that the solutions will be bounded in time. It follows that

$$\frac{\mathrm{d}^2f(t)}{\mathrm{d}t^2} + \omega^2 f(t) = 0 \tag{23}$$

$$\frac{\mathrm{d}^2\phi(x)}{\mathrm{d}t^2} + \omega^2 \frac{\rho}{E}\,\phi(x) = 0 \tag{24}$$

These have the solutions

$$f(t) = A\,\sin\omega t + B\,\cos\omega t \tag{25}$$

$$\phi(x) = C\,\sin\omega\,\sqrt{\frac{\rho}{E}}\,x + D\,\cos\omega\,\sqrt{\frac{\rho}{E}}\,x \tag{26}$$

hence for each value of ω, one has a solution of the form

$$u(x,\,t) - (A\,\sin\omega t + B\,\cos\omega t)\left(C\,\sin\omega\,\sqrt{\frac{\rho}{E}}\,x + D\,\cos\omega\,\sqrt{\frac{\rho}{E}}\,x\right) \tag{27}$$

The frequencies ω are obtained by application of the boundary conditions. For example, in the case of the clamped–free (C–F) bar shown in

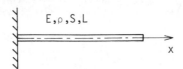

Figure 4   Clamped–free bar
in longitudinal motion

Figure 4, one must have for all instants of time

$$\phi(0) = 0$$

and

$$P(L) = ES\,\frac{\mathrm{d}\phi(x)}{\mathrm{d}x}\bigg|_{(x=L)} = 0 \tag{28}$$

The first condition requires that $D = 0$ and the second that

$$C\,\cos\omega\,\sqrt{\frac{\rho}{E}}\,L = 0 \tag{29}$$

After discarding solutions which are identically zero, this requires that

$$\cos \omega \sqrt{\frac{\rho}{E}} L = 0 \tag{30}$$

and therefore the frequencies are given by

$$\omega_n = (2n - 1)\frac{\pi}{2L} \sqrt{\frac{E}{\rho}} \qquad n = 1, 2, \dots . \tag{31}$$

To within an arbitrary constant, the modes associated with ω_n are then

$$\phi_n(x) = \sin (2n - 1)\frac{\pi x}{2L} \tag{32}$$

and the general solution for free vibration has the form

$$u(x, t) = \sum_{n=1}^{\infty} (A_n \sin \omega_n t + B_n \cos \omega_n t) \sin (2n - 1)\frac{\pi x}{2L} \tag{33}$$

The constants A_n and B_n are determined by the initial conditions; see exercise 12.

In other cases, such as clamped–clamped bars (C–C) or free–free bars (F–F), the calculation proceeds in the same fashion. All the above results are summarized in Table 1.

Table 1

C–F	$\omega_n = \dfrac{(2n-1)\pi}{2L} \sqrt{\dfrac{E}{\rho}}$	$\phi_n(x) = \sin (2n-1)\dfrac{\pi x}{2L}$	with	$n = 1, 2, \dots$
C–C	$\omega_n = \dfrac{n\pi}{L} \sqrt{\dfrac{E}{\rho}}$	$\phi_n(x) = \sin n\dfrac{\pi x}{L}$	with	$n = 1, 2, \dots$
F–F	$\omega_n = \dfrac{n\pi}{L} \sqrt{\dfrac{E}{\rho}}$	$\phi_n(x) = \cos n\dfrac{\pi x}{L}$	with	$n = 0, 1, 2, \dots$

Orthogonality relations

Using the preceeding results, equation (4) for free vibration can be written

$$\frac{d}{dx}\left(ES\frac{d\phi}{dx}\right) = -\rho S\omega^2 \phi \tag{34}$$

where, for simplicity, ϕ is written for $\phi(x)$. Equation (34) is true for each of the solution pairs: ω_i, ϕ_i; ω_j, ϕ_j. Hence,

$$\frac{d}{dx}\left(ES\frac{d\phi_i}{dx}\right) = -\rho S\omega_i^2 \phi_i \tag{35}$$

$$\frac{d}{dx}\left(ES\frac{d\phi_j}{dx}\right) = -\rho S\omega_j^2 \phi_j \tag{36}$$

If the ends of the bar are denoted by 0 and L then after multiplying (35) by ϕ_j and (36) by ϕ_i, one obtains

$$\int_0^L \phi_j \frac{d}{dx}\left(ES\frac{d\phi_i}{dx}\right) dx = -\omega_i^2 \int_0^L \rho S \phi_i \phi_j \, dx \tag{37}$$

$$\int_0^L \phi_i \frac{d}{dx}\left(ES\frac{d\phi_j}{dx}\right) dx = -\omega_j^2 \int_0^L \rho S \phi_i \phi_j \, dx \tag{38}$$

Integrating (37) and (38) by parts and assuming free ($ES \, d\phi/dx = 0$) and clamped ($\phi = 0$) boundary conditions yields the results:

$$-\int_0^L ES\frac{d\phi_i}{dx}\frac{d\phi_j}{dx} dx = -\omega_i^2 \int_0^L \rho S \phi_i \phi_j \, dx \tag{39}$$

$$-\int_0^L ES\frac{d\phi_i}{dx}\frac{d\phi_j}{dx} dx = -\omega_j^2 \int_0^L \rho S \phi_i \phi_j \, dx \tag{40}$$

Subtracting (40) from (39) gives

$$(\omega_j^2 - \omega_i^2) \int_0^L \rho S \phi_i \phi_j \, dx = 0 \tag{41}$$

and since $\omega_i \neq \omega_j$

$$\int_0^L \rho S \phi_i \phi_j \, dx = 0 \tag{42}$$

and from (39)

$$\int_0^L ES\frac{d\phi_i}{dx}\frac{d\phi_j}{dx} dx = 0 \tag{43}$$

Equations (42) and (43) are the orthogonality conditions for a continuous system deforming as a bar. Note that (42) and (43) are generalizations of equations (25) and (26) in Chapter 3.

Also note that multiplying both sides of (35) by ϕ_i and integrating from 0 to L gives

$$\int_0^L \phi_i \frac{d}{dx}\left(ES\frac{d\phi_i}{dx}\right) dx = -\omega_i^2 \int_0^L \rho S \phi_i^2 \, dx \tag{44}$$

which, on integration by parts and use of the boundary conditions, becomes

$$\omega_i^2 = \frac{\displaystyle\int_0^L ES\left(\frac{d\phi_i}{dx}\right)^2 dx}{\displaystyle\int_0^L \rho S \phi_i^2 \, dx} = \frac{k_i}{m_i} \tag{45}$$

with

$$k_i = \int_0^L ES \left(\frac{d\phi_i}{dx}\right)^2 dx \tag{46}$$

$$m_i = \int_0^L \rho S \phi_i^2 \, dx \tag{47}$$

where k_i and m_i are the ith modal stiffness and modal mass of this continuous system.

4.2.2 Bending motion

Let

$$v(x, t) = \phi(x) f(t) \tag{48}$$

and substitute in (19). This gives

$$EI \frac{d^4\phi(x)}{dx^4} f(t) + \rho S \phi(x) \frac{d^2 f(t)}{dt^2} = 0 \tag{49}$$

which can be separated into

$$\frac{EI}{\rho S} \frac{1}{\phi(x)} \frac{d^4\phi(x)}{dx^4} = -\frac{1}{f(t)} \frac{d^2 f(t)}{dt^2} = \text{const.} = \omega^2 \tag{50}$$

where, in this case, the separation constant must be set equal to $+\omega^2$. It follows that

$$\frac{d^2 f(t)}{dt^2} + \omega^2 f(t) = 0 \tag{51}$$

$$\frac{d^4\phi(x)}{dx^4} - \frac{\rho S}{EI} \omega^2 \phi(x) = 0 \tag{52}$$

The solution to (51) is

$$f(t) = A \sin \omega t + B \cos \omega t \tag{53}$$

Solutions to (52) are sought in the form e^{rx} resulting in the characteristic equations:

$$r^4 - \frac{\rho S \omega^2}{EI} = 0 \tag{54}$$

which has the roots

$$r = \beta, -\beta, j\beta, -j\beta$$

with

$$\beta = \sqrt[4]{\frac{\rho S \omega^2}{EI}} \tag{55}$$

Hence for each value of β, one has a solution of the form

$$\phi(x) = C \sin \beta x + D \cos \beta x + E \sinh \beta x + F \cosh \beta x \tag{56}$$

The frequencies ω associated with each β are determined by application of the boundary conditions. The most frequent boundary conditions for beams are

$$\begin{array}{lll} \text{Free (F):} & M = 0, & Q = 0 \\ \text{Clamped (C):} & v = 0, & \psi = 0 \\ \text{Simply-supported (S):} & v = 0, & M = 0 \end{array} \tag{57}$$

In the case of a beam which is clamped at $x = 0$ and free at $x = L$, equations (17), (56), and (57) give

$$\begin{aligned} D + F &= 0 \\ C + E &= 0 \end{aligned} \tag{58}$$

$$-C \sin \beta L - D \cos \beta L + E \sinh \beta L + F \cosh \beta L = 0$$

$$-C \cos \beta L + D \sin \beta L + E \cosh \beta L + F \sinh \beta L = 0$$

After discarding solutions which are identically zero, these equations require that the determinant of their coefficients must vanish. Completing this calculation gives

$$1 + \cos \beta L \cosh \beta L = 0 \tag{59}$$

The solutions to this equation have the form

$$\beta_n L = X_n \tag{60}$$

Accordingly, the frequencies become

$$\omega_n = \frac{X_n^2}{L^2} \sqrt{\frac{EI}{\rho S}} \tag{61}$$

The five lowest values of X_n^2 are given in Table 2 for the most common boundary conditions. Except for the case of X_1^2 for the C–F beam, all these values of X_n^2 can be easily obtained; see exercises 7 and 8. In addition, there are two rigid-body modes for the F–F beam and one rigid-body mode for the S–F beam; these zero frequencies have not been included in Table 2.

Table 2

		X_1^2	X_2^2	X_3^2	X_4^2	X_5^2
C–F	$1 + \cosh X \cos X = 0$	3.516	22.03	61.69	120.9	199.8
S–S	$\sin X = 0$	9.869	39.47	88.82	157.9	246.7
C–C F–F	$1 - \cosh X \cos X = 0$	22.37	61.67	120.9	199.8	298.5
C–S F–S	$\tan X = \tanh X$	15.41	49.96	104.2	178.2	272.0

C-F

C-S

C-C

F-F

F-S

S-S

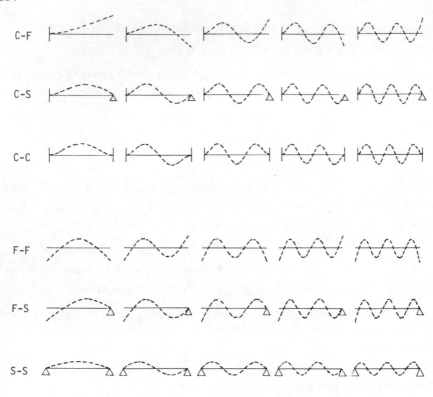

Figure 5 Five lowest bending modes of beams with various boundary conditions

The relations (56) together with (58) and (61) allow the determination of the modes. Their shapes are depicted in Figure 5.

Orthogonality relations

Starting with (18) for free vibration and again using ϕ for $\phi(x)$, one finds

$$\frac{d^2}{dx^2}\left(EI\frac{d^2\phi}{dx^2}\right) = \rho S\omega^2\phi \tag{62}$$

These equations are true for each of the solutions pairs: ω_i, ϕ_i; ω_j, ϕ_j. Hence, one has

$$\frac{d^2}{dx^2}\left(EI\frac{d^2\phi_i}{dx^2}\right) = \rho S\omega_i^2\phi_i \tag{63}$$

$$\frac{d^2}{dx^2}\left(EI\frac{d^2\phi_j}{dx^2}\right) = \rho S\omega_j^2\phi_j \tag{64}$$

Using the same procedure as was used for longitudinal motion and integrating twice by parts, one finds for each of the three boundary conditions F, C, S:

$$\int_0^L EI \frac{d^2\phi_i}{dx^2} \frac{d^2\phi_j}{dx^2} dx = \omega_i^2 \int_0^L \rho S \phi_i \phi_j \, dx \tag{65}$$

$$\int_0^L EI \frac{d^2\phi_i}{dx^2} \frac{d^2\phi_j}{dx^2} dx = \omega_j^2 \int_0^L \rho S \phi_i \phi_j \, dx \tag{66}$$

As $\omega_i \neq \omega_j$, subtracting (65) and (66) yields the orthogonality relations

$$\int_0^L \rho S \phi_i \phi_j \, dx = 0 \tag{67}$$

and from (65),

$$\int_0^L EI \frac{d^2\phi_i}{dx^2} \frac{d^2\phi_j}{dx^2} dx = 0 \tag{68}$$

In the same manner as before,

$$\omega_i^2 = \frac{k_i}{m_i} \tag{69}$$

with

$$k_i = \int_0^L EI \left(\frac{d^2\phi_i}{dx^2}\right)^2 dx \tag{70}$$

$$m_i = \int_0^L \rho S \phi_i^2 \, dx \tag{71}$$

where k_i and m_i are the ith modal stiffness and modal mass.

4.3 Approximate Methods

This section will show how the Rayleigh–Ritz method can be used to obtain estimates of frequencies and modes for continuous systems. First, the kinetic and strain energies are calculated based on a reasonable hypothesis for the system displacement. In making this hypothesis it is essential that the assumed displacement pattern satisfy the geometric boundary conditions. Then the equations of motion are deduced by application of Lagrange's equations. Finally, the frequencies and modes are obtained by using the methods presented in Chapter 3. When a single parameter function is used as a displacement hypothesis, as in Chapter 1, this procedure is Rayleigh's method.

The expressions for the kinetic energy T and the strain energy U for straight member of length L are given by the following:

1. Longitudinal motion of a bar

$$T = \frac{1}{2} \int_0^L \rho S \left(\frac{\partial u(x, t)}{\partial t} \right)^2 dx \tag{72}$$

$$U = \frac{1}{2} \int_0^L ES \left(\frac{\partial u(x, t)}{\partial x} \right)^2 dx \tag{73}$$

2. Torsional motion of a rod

$$T = \frac{1}{2} \int_0^L I_\theta \left(\frac{\partial \theta(x, t)}{\partial t} \right)^2 dx \tag{74}$$

$$U = \frac{1}{2} \int_0^L GJ \left(\frac{\partial \theta(x, t)}{\partial x} \right)^2 dx \tag{75}$$

3. Bending motion of a beam

$$T = \frac{1}{2} \int_0^L \rho S \left(\frac{\partial v(x, t)}{\partial t} \right)^2 dx \tag{76}$$

$$U = \frac{1}{2} \int_0^L EI \left(\frac{\partial^2 v(x, t)}{\partial x^2} \right)^2 dx \tag{77}$$

4.3.1 Rayleigh's method

Consider for example a clamped–free beam and assume that its lateral deflection is given by

$$v(x, t) = \left[3 \left(\frac{x}{L} \right)^2 - \left(\frac{x}{L} \right)^3 \right] p \tag{78}$$

This expression satisfies the geometric boundary conditions $v(0, t) = 0$ and $\partial v(0, t)/\partial x = \psi(0, t) = 0$, or equivalently $\phi(0) = 0$ and $\partial \phi(0)/\partial x = 0$. It also satisfies the moment boundary condition $M(0, t) = 0$, or equivalently $\partial^2 \phi(0)/\partial x^2 = 0$. The latter condition is useful but not essential. Substituting (78) in (76) and (77) gives

$$T = \frac{1}{2} \int_0^L \rho S \left[3 \left(\frac{x}{L} \right)^2 - \left(\frac{x}{L} \right)^3 \right]^2 dx p^{\circ 2}$$

$$= 0.4714 \rho S L p^{\circ 2} \tag{79}$$

$$U = \frac{1}{2} \int_0^L EI \left[\frac{6}{L^2} - \frac{6x}{L^3} \right]^2 dx p^2$$

$$= 6 \frac{EI}{L^3} p^2 \tag{80}$$

Hence the equation of motion is

$$0.9428\rho SLp^{\infty} + 12\frac{EI}{L^3}p = 0 \tag{81}$$

and the frequency is

$$\omega_1 = \frac{3.567}{L^2}\sqrt{\frac{EI}{\rho S}} \tag{82}$$

This result is 2% greater than the exact value.

4.3.2 The Rayleigh–Ritz method

Consider for example the longitudinal vibration of a bar which is clamped at $x = 0$ and free at $x = L$. It is required to estimate the first two frequencies and their corresponding modes. A four-parameter displacement hypothesis which satisfies the geometric boundary conditions is utilized; namely,

$$u(x, t) = \frac{x}{L}p_1 + \left(\frac{x}{L}\right)^2 p_2 + \left(\frac{x}{L}\right)^3 p_3 + \left(\frac{x}{L}\right)^4 p_4 \tag{83}$$

This is substituted into (72) and (73) and then Lagrange's equations are used to give the equations of motion:

$$\rho SL\begin{bmatrix} \frac{1}{3} & \frac{1}{4} & \frac{1}{5} & \frac{1}{6} \\ \frac{1}{4} & \frac{1}{5} & \frac{1}{6} & \frac{1}{7} \\ \frac{1}{5} & \frac{1}{6} & \frac{1}{7} & \frac{1}{8} \\ \frac{1}{6} & \frac{1}{7} & \frac{1}{8} & \frac{1}{9} \end{bmatrix}\begin{bmatrix} p_1^{\infty} \\ p_2^{\infty} \\ p_3^{\infty} \\ p_4^{\infty} \end{bmatrix} + \frac{ES}{L}\begin{bmatrix} 1 & 1 & 1 & 1 \\ 1 & \frac{4}{3} & \frac{3}{2} & \frac{8}{5} \\ 1 & \frac{3}{2} & \frac{9}{5} & 2 \\ 1 & \frac{8}{5} & 2 & \frac{16}{7} \end{bmatrix}\begin{bmatrix} p_1 \\ p_2 \\ p_3 \\ p_4 \end{bmatrix} = 0 \tag{84}$$

The frequencies and modes are obtained from (84):

$$\omega_1 = \frac{1.571}{L}\sqrt{\frac{E}{\rho}} \quad \text{with} \quad \phi_1 = \frac{x}{L} + 0.02782\left(\frac{x}{L}\right)^2$$
$$-0.5001\left(\frac{x}{L}\right)^3 + 0.1106\left(\frac{x}{L}\right)^4 \tag{85}$$

$$\omega_2 = \frac{4.724}{L}\sqrt{\frac{E}{\rho}} \quad \text{with} \quad \phi_2 = \frac{x}{L} - 0.6898\left(\frac{x}{L}\right)^2$$
$$-2.561\left(\frac{x}{L}\right)^3 + 2.056\left(\frac{x}{L}\right)^4 \tag{86}$$

The frequencies are accurate to within less than 1% and the modes are good estimates of the exact modes.

4.4 Response to excitation

In this section, the Rayleigh–Ritz method is used to determine the steady-state response of a continuous system to an excitating force. If, in addition, the exact modes of free vibration of the system are known, the method can be extended to permit a modal calculation of the response.

The kinetic energy and strain energy are calculated as before. The generalized forces are deduced from the expression for virtual work of external forces.

4.4.1 Rayleigh–Ritz method

Consider, as in section 4.3, the longitudinal motion of a clampled–free bar. In the present case the bar is subjected to a longitudinal force $F(t)$ at its midpoint. From (83) the displacement at the midpoint is

$$u\left(\frac{L}{2},t\right)=\tfrac{1}{2}p_1+\tfrac{1}{4}p_2+\tfrac{1}{8}p_3+\tfrac{1}{16}p_4 \tag{87}$$

and the external virtual work is therefore:

$$\delta W = F(t)\left(\frac{\delta p_1}{2}+\frac{\delta p_2}{4}+\frac{\delta p_3}{8}+\frac{\delta p_4}{16}\right) \tag{88}$$

The right-hand side of (88) gives the generalized force vector

$$[\tfrac{1}{2}F(t), \tfrac{1}{4}F(t), \tfrac{1}{8}F(t), \tfrac{1}{16}F(t)]^t \tag{89}$$

The system response can now be obtained by the methods described in Chapter 3.

4.4.2 Modal method

A bar which is clamped at each end is subjected to a longitudinal force $F(t) = F \sin \Omega t$ at its midpoint. The exact modes are used, then:

$$u(x, t) = \sin \frac{\pi x}{L} p_1 + \sin \frac{2\pi x}{L} p_2 + \sin \frac{3\pi x}{L} p_3 + \ldots \tag{90}$$

Substituting (90) into (72) and (73) and using the orthogonality relations (42) and (43), gives

$$T = \tfrac{1}{2} \sum_{n=1}^{\infty} \int_0^L \rho S \sin^2 \frac{n\pi x}{L} \, dx \, p_n{}^{\circ 2} = \frac{\rho SL}{4} \sum_{n=1}^{\infty} p_n{}^{\circ 2} \tag{91}$$

$$U = \tfrac{1}{2} \sum_{n=1}^{\infty} \int_0^L ES \left(\frac{n\pi}{L}\right)^2 \cos^2 \frac{n\pi x}{L} \, dx \, p_n^2 = \frac{\pi^2}{4} \frac{ES}{L} \sum_{n=1}^{\infty} n^2 p_n^2 \tag{92}$$

The virtual work of external force $F(t)$ is

$$\delta W = F \sin \Omega t \left(\sin \frac{\pi}{2} \delta p_1 + \sin \pi \, \delta p_2 + \sin \frac{3\pi}{2} \delta p_3 + \ldots\right) \tag{93}$$

The use of (91), (92), and (93) in Lagrange's equations gives a set of uncoupled modal equations for the parameters $p_i(t)$; namely:

$$\frac{\rho SL}{2} \overset{\infty}{p_1} + \frac{\pi^2}{2} \frac{ES}{L} p_1 = F \sin \Omega t$$

$$\frac{\rho SL}{2} \overset{\infty}{p_2} + \frac{4\pi^2}{2} \frac{ES}{L} p_2 = 0 \tag{94}$$

$$\frac{\rho SL}{2} \overset{\infty}{p_3} + \frac{9\pi^2}{2} \frac{ES}{L} p_3 = -F \sin \Omega t$$

$$\vdots$$

For i even, the generalized forces are zero, hence the steady-state solutions for $i = 2, 4, \ldots$ are $p_i = 0$. Using the steady-state solutions to (94), for i odd, and substituting the results into (90), gives

$$u(x, t) = \frac{\sin \dfrac{\pi x}{L} F \sin \Omega t}{\dfrac{\pi^2 ES}{2L} - \Omega^2 \dfrac{\rho SL}{2}} - \frac{\sin \dfrac{3\pi x}{L} F \sin \Omega t}{\dfrac{9\pi^2}{2} \dfrac{ES}{L} - \Omega^2 \dfrac{\rho SL}{2}} \tag{95}$$

It is evident that for $\Omega = (\pi/L)\sqrt{(E/\rho)}, (3\pi/L)\sqrt{(E/\rho)}, \ldots$, the system is resonant. These resonant frequencies correspond to the frequencies of free vibration identified in Table 1 for a C–C bar.

The longitudinal force in the bar can be obtained by using (95) and (3).

4.5   Kinetic and Strain Energies of Plates

The formulation of the dynamic behavior of a plate in bending can be done in an analogous manner to that used for the beam in bending. However, since a plate is thin in only one dimension while a beam is thin in two dimensions, the plate variables are a function of two space coordinates as well as time. This section will present only a brief introduction to the dynamic theory of plates and will be limited to a consideration of their kinetic and strain energies.

Let the xy plane be the middle plane of the plate so that z is the normal distance from the middle plane to a point in the plate. Also the plate is assumed thin and second order effects, i.e. rotatory inertia and shear deformation, are neglected. The classical strain–displacement relations for a plate are

$$\varepsilon_x = -z \frac{\partial^2 w}{\partial x^2}$$

$$\varepsilon_y = -z \frac{\partial^2 w}{\partial y^2} \tag{96}$$

$$\gamma_{xy} = -2z \frac{\partial^2 w}{\partial x \, \partial y}$$

The stress–strain relations are

$$\sigma = \begin{bmatrix} \sigma_x \\ \sigma_y \\ \tau_{xy} \end{bmatrix} = \frac{E}{1-\nu^2} \begin{bmatrix} 1 & \nu & 0 \\ \nu & 1 & 0 \\ 0 & 0 & \dfrac{1-\nu}{2} \end{bmatrix} \begin{bmatrix} \varepsilon_x \\ \varepsilon_y \\ \gamma_{xy} \end{bmatrix} = D\varepsilon \tag{97}$$

where ν is Poisson's ratio. The general expression for strain energy is

$$U = \tfrac{1}{2} \int_\tau \varepsilon' \sigma \, d\tau \tag{98}$$

where τ is the volume. Substituting (96) and (97) into (98) gives

$$U = \int_\tau \frac{E}{2(1-\nu^2)} \begin{bmatrix} z\dfrac{\partial^2 w}{\partial x^2} \\ z\dfrac{\partial^2 w}{\partial y^2} \\ 2z\dfrac{\partial^2 w}{\partial x\,\partial y} \end{bmatrix}^t \begin{bmatrix} 1 & \nu & 0 \\ \nu & 1 & 0 \\ 0 & 0 & \dfrac{1-\nu}{2} \end{bmatrix} \begin{bmatrix} z\dfrac{\partial^2 w}{\partial x^2} \\ z\dfrac{\partial^2 w}{\partial y^2} \\ 2z\dfrac{\partial^2 w}{\partial x\,\partial y} \end{bmatrix} d\tau \tag{99}$$

If the thickness of the plate, h, is constant, this becomes

$$U = \frac{D}{2} \int_S \left[\left(\frac{\partial^2 w}{\partial x^2}\right)^2 + \left(\frac{\partial^2 w}{\partial y^2}\right)^2 + 2\nu\left(\frac{\partial^2 w}{\partial x^2}\right)\left(\frac{\partial^2 w}{\partial y^2}\right) + 2(1-\nu)\left(\frac{\partial^2 w}{\partial x\,\partial y}\right)^2 \right] dx\,dy \tag{100}$$

with S being the surface and D being the bending stiffness of the plate:

$$D = \frac{Eh^3}{12(1-\nu^2)} \tag{101}$$

The expression for kinetic energy is

$$T = \frac{1}{2} \int_\tau \rho \left(\frac{\partial w}{\partial t}\right)^2 d\tau \tag{102}$$

which for constant thickness becomes

$$T = \frac{\rho h}{2} \int_S \left(\frac{\partial w}{\partial t}\right)^2 dx\,dy \tag{103}$$

For circular plates, U and T are better expressed in polar coordinates. If the plate thickness is constant, these expressions are

$$U = \frac{D}{2} \int_S \left[\left(\frac{\partial^2 w}{\partial r^2} + \frac{1}{r}\frac{\partial w}{\partial r} + \frac{1}{r^2}\frac{\partial^2 w}{\partial \theta^2}\right)^2 - 2(1-\nu)\frac{\partial^2 w}{\partial r^2}\left(\frac{1}{r}\frac{\partial w}{\partial r} + \frac{1}{r^2}\frac{\partial^2 w}{\partial \theta^2}\right) \right.$$
$$\left. + 2(1-\nu)\left(\frac{\partial}{\partial r}\left(\frac{1}{r}\frac{\partial w}{\partial \theta}\right)\right)^2 \right] r\, d\theta\, dr \tag{104}$$

$$T = \frac{\rho h}{2} \int_S \left(\frac{\partial w}{\partial t}\right)^2 r \, d\theta \, dr \tag{105}$$

4.6 Frequencies and Modes for Plates, Response to Excitation

The procedures are the same as used for bars and beams. They will be illustrated by three examples: two dealing with the calculation of frequency, one by Rayleigh's method and the other by the Rayleigh–Ritz method, and one dealing with the determination of steady-state response to a concentrated force.

4.6.1 Rayleigh's method

The problem is to calculate the lowest frequency of a circular plate undergoing axisymmetric vibration. The plate is fixed at its center point and free at its circumference. Consider the displacement hypothesis:

$$\dot{w}(r, t) = \left[3\left(\frac{r}{R}\right)^2 - \left(\frac{r}{R}\right)^3\right] p \tag{106}$$

where symmetry requires that w be a spatial function of r only. Equation (106) satisfies the geometric boundary conditions at the central support point. Since the displacement is not a function of θ, equations (104) and (105) become

$$U = \frac{D}{2} \int_0^R \left[\left(\frac{\partial^2 w}{\partial r^2}\right)^2 + \frac{1}{r^2}\left(\frac{\partial w}{\partial r}\right)^2 + \frac{2\nu}{r}\frac{\partial w}{\partial r}\frac{\partial^2 w}{\partial r^2}\right] 2\pi r \, dr \tag{107}$$

$$T = \frac{\rho h}{2} \int_0^R \left(\frac{\partial w}{\partial t}\right)^2 2\pi r \, dr \tag{108}$$

Substituting (106) in (107) and (108) gives

$$U = \frac{\pi D}{R^2}(11.25 + 9\nu)p^2 \tag{109}$$

$$T = 2.412 \rho h R^2 p^{\circ 2} \tag{110}$$

from which the frequency is found to be

$$\omega = \frac{1}{R^2}\sqrt{\frac{(11.25 + 9\nu)D\pi}{2.412\rho h}}$$

$$= \frac{h}{R^2}\sqrt{\frac{(11.25 + 9\nu)E\pi}{28.94(1 - \nu^2)\rho}} \tag{111}$$

This result is 15% higher than the exact value.

4.6.2 Rayleigh–Ritz method

A rectangular plate of constant thickness is simply supported, i.e. no lateral displacement and no bending moment, along all its edges and has a

112

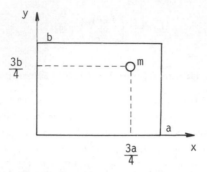

Figure 6 Rectangular plate simply supported along all four edges with a concentrated mass at $(3a/4, 3b/4)$

concentrated mass of magnitude $m = \rho abh/10$ at the point $x = 3a/4$, $y = 3b/4$; see Figure 6. The displacement hypothesis is

$$w(x, y, t) = \sum_{r=1}^{\infty} \sum_{s=1}^{\infty} \sin \frac{r\pi x}{a} \sin \frac{s\pi y}{b} p_{rs} \tag{112}$$

which satisfies the geometric boundary conditions at the plate edges (i.e. $w = 0$) as well as the force boundary conditions at the edge: $\partial^2 w/\partial x^2 = 0$ for $x = 0$, a; $\partial^2 w/\partial y^2 = 0$ for $y = 0, b$. The latter conditions are, of course, not essential.

Substituting (112) into (100) and (103) and taking advantage of the orthogonality of sine functions, one finds

$$U = \frac{\pi^2 abD}{8} \sum_{r=1}^{\infty} \sum_{s=1}^{\infty} \left(\frac{r^2}{a^2} + \frac{s^2}{b^2}\right)^2 p_{rs}^2 \tag{113}$$

$$T = \frac{abh\rho}{8} \sum_{r=1}^{\infty} \sum_{s=1}^{\infty} \overset{\circ 2}{p_{rs}} \tag{114}$$

Expanding (114) for $r = 1, 2$ and $s = 1, 2$ gives

$$U = \frac{\pi^4 abD}{8} \left[\left(\frac{1}{a^2} + \frac{1}{b^2}\right)^2 p_{11}^2 + \left(\frac{4}{a^2} + \frac{1}{b^2}\right)^2 p_{21}^2 \right.$$
$$\left. + \left(\frac{1}{a^2} + \frac{4}{b^2}\right)^2 p_{12}^2 + \left(\frac{4}{a^2} + \frac{4}{b^2}\right)^2 p_{22}^2 \right] \tag{115}$$

$$T = \frac{abh\rho}{8} (\overset{\circ 2}{p_{11}} + \overset{\circ 2}{p_{21}} + \overset{\circ 2}{p_{12}} + \overset{\circ 2}{p_{22}}) \tag{116}$$

The kinetic energy of the concentrated mass must be added to (116). This increment in kinetic energy is given by:

$$T_m = \frac{abh\rho}{20} \left(0.5\overset{\circ}{p_{11}} - \frac{\sqrt{2}}{2}\overset{\circ}{p_{21}} - \frac{\sqrt{2}}{2}\overset{\circ}{p_{12}} + \overset{\circ}{p_{22}}\right)^2 \tag{117}$$

the above energy expressions will be simplified by considering the special case of $b = a/2$. The total kinetic energy, i.e. the sum of (116) and (117), is

$$T = \frac{a^2 h \rho}{16} (1.1 \dot{p}_{11}^2 + 1.2 \dot{p}_{21}^2 + 1.2 \dot{p}_{12}^2 + 1.4 \dot{p}_{22}^2$$

$$- 0.2828 \dot{p}_{11} \dot{p}_{21} - 0.2828 \dot{p}_{11} \dot{p}_{12} + 0.4 \dot{p}_{11} \dot{p}_{22}$$

$$+ 0.4 \dot{p}_{21} \dot{p}_{12} - 0.5657 \dot{p}_{21} \dot{p}_{22} - 0.5657 \dot{p}_{12} \dot{p}_{22}) \tag{118}$$

and the strain energy is

$$U = \frac{\pi^4 D}{16 a^2} (25 p_{11}^2 + 64 p_{21}^2 + 289 p_{12}^2 + 400 p_{22}^2) \tag{119}$$

Lagrange's equations result in the equations of motion:

$$a^2 \rho h \begin{bmatrix} 1.1 & -0.1414 & -0.1414 & 0.2 \\ -0.1414 & 1.2 & 0.2 & -0.2828 \\ -0.1414 & 0.2 & 1.2 & -0.2828 \\ 0.2 & -0.2828 & -0.2828 & 1.4 \end{bmatrix} \begin{bmatrix} \ddot{p}_{11} \\ \ddot{p}_{21} \\ \ddot{p}_{12} \\ \ddot{p}_{22} \end{bmatrix}$$

$$+ \frac{\pi^4 D}{a^2} \begin{bmatrix} 25 & & & \text{zero} \\ & 64 & & \\ & & 289 & \\ \text{zero} & & & 400 \end{bmatrix} \begin{bmatrix} p_{11} \\ p_{21} \\ p_{12} \\ p_{22} \end{bmatrix} = 0 \tag{120}$$

and the frequencies and modes can be determined with the aid of program no. 5 of Chapter 7. The results are

$$\omega_1 = 4.730 \frac{\pi^2 h}{a^2} \sqrt{\frac{E}{12(1 - \nu^2)\rho}} \tag{121}$$

$$\phi_1(x, y) = \sin \frac{\pi x}{a} \sin \frac{2\pi y}{a} - 0.0892 \sin \frac{2\pi x}{a} \sin \frac{2\pi y}{a}$$

$$- 0.0139 \sin \frac{\pi x}{a} \sin \frac{4\pi y}{a} + 0.0139 \sin \frac{2\pi x}{a} \sin \frac{4\pi y}{a} \tag{122}$$

$$\omega_2 = 7.347 \frac{\pi^2 h}{a^2} \sqrt{\frac{E}{12(1 - \nu^2)\rho}} \tag{123}$$

$$\phi_2(x, y) = \sin \frac{\pi x}{a} \sin \frac{2\pi y}{a} + 4.092 \sin \frac{2\pi x}{a} \sin \frac{2\pi y}{a}$$

$$+ 0.1744 \sin \frac{\pi x}{a} \sin \frac{4\pi y}{a} - 0.1675 \sin \frac{2\pi x}{a} \sin \frac{4\pi y}{a} \tag{124}$$

4.6.3 Steady-state response

The steady-state response of the plate just studied will be determined for an excitation force $F(t)$ at the point $x = 3a/4$, $y = 3a/8$. In order to simplify the calculation, the concentrated mass will be removed. The displacement function will then give uncoupled equations of motion.

The expressions for kinetic and strain energies can be obtained by setting $b = a/2$ in (115) and (116). The generalized forces are deduced from the expression for the external virtual work of $F(t)$:

$$\delta W = F(t) \left[\sum_{r=1}^{\infty} \sum_{s=1}^{\infty} \left(\sin \frac{3\pi r}{4} \sin \frac{3\pi s}{4} \right) \delta p_{rs} \right] \tag{125}$$

which implies that the generalized force corresponding to each coordinate p_{rs} is

$$\sin \frac{3\pi r}{4} \sin \frac{3\pi s}{4} F(t) \tag{126}$$

Application of Lagrange's equation gives the uncoupled equations of motion:

$$\frac{a^2 h\rho}{8} p_{rs}^{\circ\circ} + \frac{\pi^4 a^2 D}{8} \left(\frac{r^2}{a^2} + \frac{4s^2}{a^2} \right) p_{rs} = \sin \frac{3\pi r}{4} \sin \frac{3\pi s}{4} F(t) \tag{127}$$

from which p_{rs} can be obtained and then $w(x, y, t)$ can be found from (112).

4.7 Kinetic and Strain Energies of Rotor Elements

The major elements which constitute rotating machinery are disks, shafts, and bearings. This section will present a derivation of the kinetic energy of a rigid rotating disk and the strain energy of a flexible rotating shaft. The kinetic energy of the shaft is not considered but its expression could be obtained from an extension of the kinetic energy expression of the disk. Also in the applications which are presented in the exercises, the bearings are assumed rigid.

Figure 7 shows the frames of reference which are used in the study of rotors. The axes XYZ are an inertial frame while the axes xyz are fixed to the disk. The xyz system is related to the XYZ system through a set of the three angles θ, ψ, ϕ. To achieve the orientation of the disk, one first rotates an amount ψ around the Z axis; then an amount θ around the new x-axis, denoted by x_1; and lastly an amount ϕ around the final y-axis.

4.7.1 Kinetic energy of the disk

With the aid of Figure 7, one can write the instantaneous angular velocity vector of the xyz frame as

$$\vec{\omega} = \psi^{\circ} \vec{Z} + \theta^{\circ} \vec{x}_1 + \phi^{\circ} \vec{y} \tag{128}$$

where $\vec{Z}$, $\vec{x}_1$, $\vec{y}$ are unit vectors. The angular velocity of the disk and shaft is $\phi^{\circ} = \Omega$. After some algebra, the components of $\vec{\omega}$ in the xyz coordinate

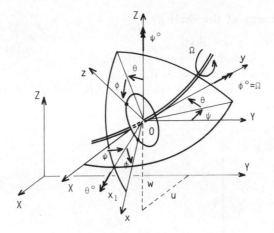

Figure 7   Reference frames for a disk on a
rotating, flexible shaft

system can be written as

$$\vec{\omega} = \begin{bmatrix} \omega_x \\ \omega_y \\ \omega_z \end{bmatrix} = \begin{bmatrix} -\psi^\circ \cos\theta \sin\phi + \theta^\circ \cos\phi \\ \phi^\circ + \psi^\circ \sin\theta \\ \psi^\circ \cos\theta \cos\phi + \theta^\circ \sin\phi \end{bmatrix} \tag{129}$$

Let u and w in Figure 7 designate the coordinates of the disk center of mass, which is assumed to be on the axis of the shaft. Then the kinetic energy of the disk is given by

$$T_D = \tfrac{1}{2} M_D (u^{\circ 2} + w^{\circ 2}) + \tfrac{1}{2}(I_{Dx}\omega_x^2 + I_{Dy}\omega_y^2 + I_{Dz}\omega_z^2) \tag{130}$$

where I_{Dx}, I_{Dy}, and I_{Dz} are the mass moments of inertia of the disk with respect to the xyz axes, and M_D is the mass of the disk. Equation (130) is general but in our case $\phi^\circ = \Omega$, and θ and ψ are small. If, in addition, $I_{Dx} = I_{Dz}$, (129) and (130) give

$$T_D = \tfrac{1}{2} M_D (u^{\circ 2} + w^{\circ 2}) + \tfrac{1}{2} I_{Dx}(\theta^{\circ 2} + \psi^{\circ 2}) + \tfrac{1}{2} I_{Dy}(\Omega^2 + 2\Omega\psi^\circ\theta) \tag{131}$$

This provides a set of generalized coordinates

$$q = (u, w, \theta, \psi)^t \tag{132}$$

which, when used in Lagrange's equations, give

$$\begin{bmatrix} M & 0 & 0 & 0 \\ 0 & M & 0 & 0 \\ 0 & 0 & I_{Dx} & 0 \\ 0 & 0 & 0 & I_{Dx} \end{bmatrix} \begin{bmatrix} u^{\circ\circ} \\ w^{\circ\circ} \\ \theta^{\circ\circ} \\ \psi^{\circ\circ} \end{bmatrix} + \Omega \begin{bmatrix} 0 & 0 & 0 & 0 \\ 0 & 0 & 0 & 0 \\ 0 & 0 & 0 & -I_{Dy} \\ 0 & 0 & I_{Dy} & 0 \end{bmatrix} \begin{bmatrix} u^\circ \\ w^\circ \\ \theta^\circ \\ \psi^\circ \end{bmatrix} \tag{133}$$

The two matrices in (133) are the mass matrix and the so-called Coriolis matrix.

4.7.2 Strain energy of the shaft

We use the following notation:

E,   Young's modulus of the shaft material
ε, σ,   the strain and stress in the shaft
u^*, v^*, w^*, the displacements of the geometric center of the shaft with respect of the xyz axes (see Figure 8).

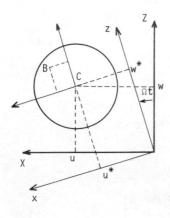

Figure 8   Coordinates of geometric center (C) and arbitrary point (B) of the shaft

The longitudinal displacement v^* is assumed to vanish. If the second-order terms are included, the longitudinal strain of a point B can be shown to be

$$\varepsilon = -z\frac{\partial^2 w^*}{\partial y^2} - x\frac{\partial^2 u^*}{\partial y^2} + \frac{1}{2}\left(\frac{\partial u^*}{\partial y}\right)^2 + \frac{1}{2}\left(\frac{\partial w^*}{\partial y}\right)^2 \qquad (134)$$

or

$$\varepsilon = \varepsilon_1 + \varepsilon_{nl} \qquad (135)$$

where ε_1 contains the linear terms and ε_{nl} the nonlinear terms of (134). The strain energy is

$$U_1 = \frac{1}{2}\int_\tau \varepsilon' \sigma \, d\tau = \tfrac{1}{2}E\int_\tau (\varepsilon_1^2 + 2\varepsilon_{nl}\varepsilon_1 + \varepsilon_{nl}^2) \, d\tau \qquad (136)$$

The symmetry of the shaft cross-section with respect to x and z result in

$$\int_\tau \varepsilon_{nl}\varepsilon_1 \, d\tau = 0$$

The term $\int_\tau \varepsilon_{nl}^2 \, d\tau$ is of the second order and neglected. The strain energy is then

$$U = \tfrac{1}{2}E\int_\tau \left[-z\frac{\partial^2 w^*}{\partial y^2} - x\frac{\partial^2 u^*}{\partial y^2}\right]^2 d\tau \qquad (137)$$

or

$$U_1 = \tfrac{1}{2}E \int_0^L I_x \left(\frac{\partial^2 w^*}{\partial y^2}\right)^2 dy + \tfrac{1}{2}E \int_0^L I_z \left(\frac{\partial^2 u^*}{\partial y^2}\right) dy \qquad (138)$$

where

$$I_x = \int_S z^2 \, dx \, dz$$

$$I_z = \int_S x^2 \, dx \, dz \qquad (139)$$

are the area moments of inertia of the shaft.

If the shaft is subjected to a constant axial force F_0, there is a second contribution to the strain energy of the shaft given by

$$U_2 = \int_\tau \left(\frac{F_0}{S}\right)[\varepsilon_1 + \varepsilon_{nl}] \, d\tau \qquad (140)$$

Because of symmetry, the integral of ε_1 over the shaft cross-sectional area will vanish, and (140) becomes

$$U_2 = \tfrac{1}{2}F_0 \int_0^L \left[\left(\frac{\partial w^*}{\partial y}\right)^2 + \left(\frac{\partial u^*}{\partial y}\right)^2\right] dy \qquad (141)$$

The combined strain energy $U_1 + U_2$ is then

$$U = \frac{1}{2}\int_0^L EI_x \left(\frac{\partial^2 w^*}{\partial y^2}\right)^2 dy + \frac{1}{2}\int_0^L EI_z \left(\frac{\partial^2 u^*}{\partial y^2}\right)^2 dy + \tfrac{1}{2}F_0 \int_0^L \left(\frac{\partial w^*}{\partial y}\right)^2 dy$$

$$+ \tfrac{1}{2}F_0 \int_0^L \left(\frac{\partial u^*}{\partial y}\right)^2 dy \qquad (142)$$

Using

$$w^* = w \cos \Omega t + u \sin \Omega t$$
$$u^* = -w \sin \Omega t + u \cos \Omega t \qquad (143)$$

which can be deduced from Figure 8, equation (142) is written

$$U = \tfrac{1}{2}E \int_0^L \left[I_x \left(\cos \Omega t \frac{\partial^2 w}{\partial y^2} + \sin \Omega t \frac{\partial^2 u}{\partial y^2}\right)^2 \right.$$

$$\left. + I_z \left(-\sin \Omega t \frac{\partial^2 w}{\partial y^2} + \cos \Omega t \frac{\partial^2 u}{\partial y^2}\right)^2 \right] dy$$

$$+ \tfrac{1}{2}F_0 \int_0^L \left[\left(\frac{\partial w}{\partial y}\right)^2 + \left(\frac{\partial u}{\partial y}\right)^2\right] dy \qquad (144)$$

Finally, for the most common case of a symmetric shaft, i.e. $I_x = I_z = I$, the strain energy becomes

$$U = \frac{EI}{2} \int_0^L \left[\left(\frac{\partial^2 w}{\partial y^2} \right)^2 + \left(\frac{\partial^2 u}{\partial y^2} \right)^2 \right] dy + \frac{F_0}{2} \int_0^L \left[\left(\frac{\partial w}{\partial y} \right)^2 + \left(\frac{\partial u}{\partial y} \right)^2 \right] dy \quad (145)$$

Equations (131) and (145) allow the derivation of equations of motion for the system of Figure 7 via Lagrange's equations. Previous methods can be used to solve these equations of motion and to assist in the design of a rotating rotor–disk system. The determination of system frequencies and critical speeds due to unbalance is left to the exercises.

4.8 Exercises

Exercise 1: *Bar in longitudinal motion. The bar is fixed at one end and has a spring of stiffness k at the other end; see Figure 9. Calculate the first two frequencies and their corresponding modes for $k = ES/L$.*

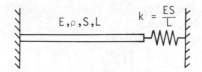

Figure 9

Since the bar is fixed at the origin, equation (26) implies

$$\phi(x) = \sin \omega \sqrt{\frac{\rho}{E}} x$$

and as

$$P(L) = ES \frac{d\phi(x)}{dx} \bigg|_{x=L} = -k\phi(L) = -\frac{ES}{L} \phi(L)$$

it follows that

$$ES\omega \sqrt{\frac{\rho}{E}} \cos \omega \sqrt{\frac{\rho}{E}} L = -\frac{ES}{L} \sin \omega \sqrt{\frac{\rho}{E}} L$$

Let

$$X = \omega \sqrt{\frac{\rho}{E}} L$$

then the above becomes

$$\tan X = -X$$

The first two lowest values of X which satisfy this transcendental equation are

$$\omega_1 = \frac{2.029}{L} \sqrt{\frac{E}{\rho}}; \qquad \omega_2 = \frac{4.913}{L} \sqrt{\frac{E}{\rho}}$$

and the modes are

$$\phi_1(x) = \sin 2.029 \frac{x}{L}; \qquad \phi_2(x) = \sin 4.913 \frac{x}{L}$$

Exercise 2: *Consider a bar which is fixed at one end and free at the other. At the free end it has a concentrated mass M. Find the equation for the frequencies.*

Proceeding as in exercise 1, one finds

$$\tan \omega \sqrt{\frac{\rho}{E}} L = \frac{ES}{M\omega} \sqrt{\frac{\rho}{E}}$$

Exercise 3: *Consider the bar shown in Figure 10 which is fixed at one end and free at the other and undergoes longitudinal motion. Find the relation which allows the calculation of the frequencies for this system.*

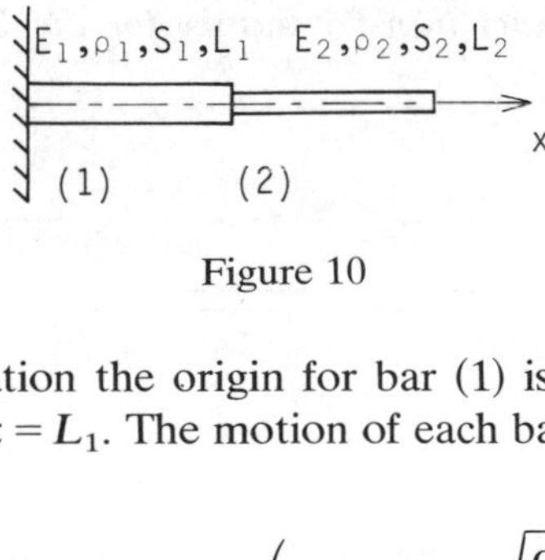

Figure 10

To simplify the equation the origin for bar (1) is taken at $x = 0$ and the origin for bar (2) is at $x = L_1$. The motion of each bar is defined by equation (27):

$$u_1(x, t) = (A_1 \sin \omega_{(1)}t + B_1 \cos \omega_{(1)}t)\left(C_1 \sin \omega_{(1)} \sqrt{\frac{\rho_1}{E_1}} x + D_1 \cos \omega_{(1)} \sqrt{\frac{\rho_1}{E_1}} x\right)$$

$$u_2(x, t) = (A_2 \sin \omega_{(2)}t + B_2 \cos \omega_{(2)}t)\left(C_2 \sin \omega_{(2)} \sqrt{\frac{\rho_2}{E_2}} x + D_2 \cos \omega_{(2)} \sqrt{\frac{\rho_2}{E_2}} x\right)$$

Since bar (1) is fixed at $x = 0$, $u_1(0, t) = 0$ and $D_1 = 0$. Continuity of displacement at the junction of the two bars is assumed for all t and hence requires

$$A_1 = A_2 = A; \qquad B_1 = B_2 = B; \qquad \omega_{(1)} = \omega_{(2)} = \omega$$

$$D_2 = C_1 \sin \omega \sqrt{\frac{\rho_1}{E_1}} L_1$$

Equality of the forces at the junction gives

$$C_1\left[E_1 S_1 \omega \sqrt{\frac{\rho_1}{E_1}} \cos \omega \sqrt{\frac{\rho_1}{E_1}} L_1\right] = C_2\left[E_2 S_2 \omega \sqrt{\frac{\rho_2}{E_2}}\right]$$

The force must vanish at the free end, which implies

$$C_2 \cos \omega \sqrt{\frac{\rho_2}{E_2}} L_2 - D_2 \sin \omega \sqrt{\frac{\rho_2}{E_2}} L_2 = 0$$

Putting these three equations in matrix form gives

$$\begin{bmatrix} \sin \gamma_1 L_1 & 0 & -1 \\ E_1 S_1 \gamma_1 \cos \gamma_1 L_1 & -E_2 S_2 \gamma_2 & 0 \\ 0 & \cos \gamma_2 L_2 & -\sin \gamma_2 L_2 \end{bmatrix}\begin{bmatrix} C_1 \\ C_2 \\ D_2 \end{bmatrix} = \begin{bmatrix} 0 \\ 0 \\ 0 \end{bmatrix}$$

with

$$\gamma_1 = \omega \sqrt{\frac{\rho_1}{E_1}}; \qquad \gamma_2 = \omega \sqrt{\frac{\rho_2}{E_2}}$$

The frequencies are then determined by setting the determinant of the coefficients equal to zero. This gives

$$E_2 S_2 \gamma_2 \sin \gamma_1 L_1 \sin \gamma_2 L_2 - E_1 S_1 \gamma_1 \cos \gamma_1 L_1 \cos \gamma_2 L_2 = 0$$

Exercise 4: *Find the exact transfer matrix for the longitudinal motion of a uniform bar; see Figure* 11.

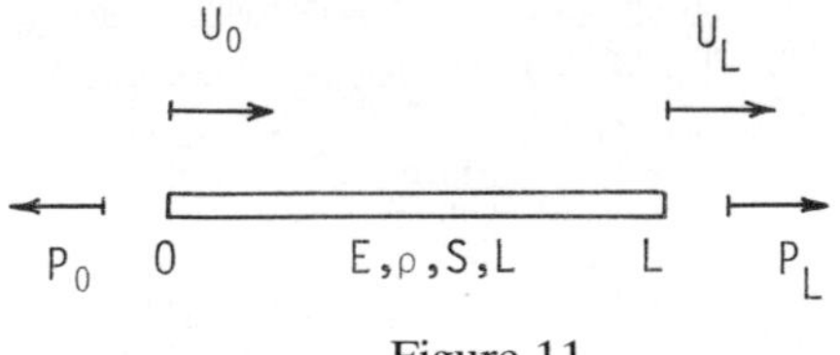

Figure 11

Referring to (26) and (3) and using $u(x, t) = U(x) f(t)$ gives

$$U(x) = C \sin \omega \sqrt{\frac{\rho}{E}} x + D \cos \omega \sqrt{\frac{\rho}{E}} x$$

$$P(x) = ES \frac{dU(x)}{dx}$$

$$= \omega \sqrt{\frac{\rho}{E}} ES\left(C \cos \omega \sqrt{\frac{\rho}{E}} x - D \sin \omega \sqrt{\frac{\rho}{E}} x\right)$$

Since at

$$x = 0, \qquad U(x) = U_0 \qquad \text{and} \qquad P(x) = P_0$$
$$x = L, \qquad U(L) = U_L \qquad \text{and} \qquad P(L) = P_L$$

then

$$U_0 = D \quad \text{and} \quad P_0 = \omega \sqrt{\frac{\rho}{E}} \, ESC$$

Let

$$\gamma = \omega \sqrt{\frac{\rho}{E}}$$

then

$$U(x) = \frac{\sin \gamma x}{ES\gamma} F_0 + \cos \gamma x U_0$$

and

$$\begin{bmatrix} P_L \\ U_L \end{bmatrix} = \begin{bmatrix} \cos \gamma L & -ES\gamma \sin \gamma L \\ \dfrac{\sin \gamma L}{ES\gamma} & \cos \gamma L \end{bmatrix} \begin{bmatrix} P_0 \\ U_0 \end{bmatrix}$$

Exercise 5: *Consider the same system as in exercise 1. With the aid of the resultant of exercise 4, rederive the frequency equation for the system; see Figure* 12.

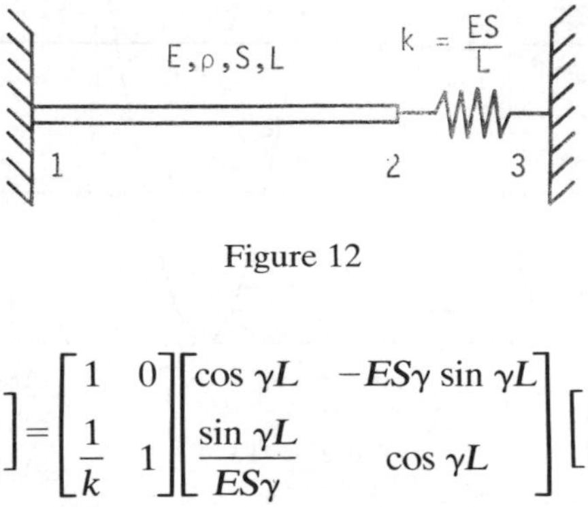

Figure 12

$$\begin{bmatrix} P_3 \\ U_3 \end{bmatrix} = \begin{bmatrix} 1 & 0 \\ \dfrac{1}{k} & 1 \end{bmatrix} \begin{bmatrix} \cos \gamma L & -ES\gamma \sin \gamma L \\ \dfrac{\sin \gamma L}{ES\gamma} & \cos \gamma L \end{bmatrix} \begin{bmatrix} P_1 \\ U_1 \end{bmatrix}$$

$$U_3 = \left(\frac{\cos \gamma L}{k} + \frac{\sin \gamma L}{ES\gamma} \right) P_1 + \left(\cos \gamma L - \frac{ES\gamma}{k} \sin \gamma L \right) U_1$$

The frequency equation is obtained from the boundary conditions $U_3 = 0$ and $U_1 = 0$ which result in

$$\frac{\cos \gamma L}{k} + \frac{\sin \gamma L}{ES\gamma} = 0$$

Since $k = ES/L$ and setting $X = \gamma L = \omega \sqrt{\rho/E}\, L$, one obtains

$$\tan X = -X$$

Exercise 6: *Find the orthogonality relationships for the system of exercise* 2.

The solutions are:

$$\int_0^L ES \frac{d\phi_i(x)}{dx} \frac{d\phi_j(x)}{dx} = 0$$

$$\int_0^L \rho S\phi_i(x)\phi_j(x) + M\phi_i(L)\phi_j(L) = 0$$

Exercise 7: *Plot the two functions* cos X *and* $-1/\cosh X$ *and determine their points of intersection. Relate these to the three lowest frequencies of a clamped free beam.*

Let X_1, X_2, X_3 be the three lowest points of intersection as shown in the Figure 13. Notice that the values of $X_2, X_3, \ldots$, rapidly approach the zeros

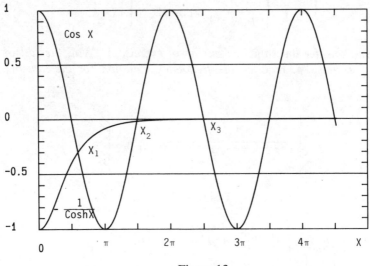

Figure 13

of the function cos X so that

$$X_2 \simeq \frac{3\pi}{2}, \qquad X_3 \simeq \frac{5\pi}{2}$$

and

$$X_2^2 \simeq 22.21, \qquad X_3^2 \simeq 61.69$$

Therefore only X_1 has to be calculated from the transcendental equation for frequency.

Exercise 8: *Show that the transcendental frequency formulas for C–C, F–F, C–S, F–S beams given in Table 2 of section 4.2.2 can be simplified as in exercise 7.*

Neglecting the case of zero frequency, which is mentioned in the text, one can show that seeking graphical solutions to the transcendental equations

$$\cos X = \frac{1}{\cosh X}$$

and

$$\tan X = \tanh X$$

yields the results that:

(1) for C–C and F–F beams the values X_i^2 can be found from $\cos X = 0$ and these values are

$$X_1^2 = \left(\frac{3\pi}{2}\right)^2 = 22.21, \qquad X_2^2 = \left(\frac{5\pi}{2}\right)^2 = 61.69, \ldots$$

(2) for C–S and F–S beams, the values X_i^2 can be found from $\tan X = 1$ with $X > \pi/4$ and these values are

$$x_1^2 = \left(\frac{\pi}{4} + \pi\right)^2 = 15.42, \qquad x_2^2 = \left(\frac{\pi}{4} + 2\pi\right)^2 = 49.96, \ldots$$

Exercise 9: *Consider a beam which is simply supported at each end. Find the expression for the modes and determine the corresponding kinetic and strain energies, modal masses and stiffnesses.*

The solution is

$$\phi_n(x) = \sin\frac{n\pi x}{L}$$

Using (76) and (77),

$$T_n = \frac{\rho SL}{4}\, p_n^{\circ 2}(t); \qquad U_n = \frac{n^4\pi^4}{4L^3}\, EI p_n^2(t)$$

and from (70) and (71),

$$m_n = \frac{\rho SL}{2}; \qquad k_n = \frac{n^4\pi^4 EI}{2L^3}$$

Exercise 10: *Find the exact transfer matrix for a beam in bending using the same procedure as in exercise 4.*

The solution is

$$
\begin{bmatrix} Q_L \\ M_L \\ \psi_L \\ V_L \end{bmatrix} =
\begin{bmatrix}
\dfrac{\text{ch }\beta L + \cos\beta L}{2} & \dfrac{\beta(\text{sh }\beta L - \sin\beta L)}{2} \\[2mm]
\dfrac{\text{sh }\beta L + \sin\beta L}{2\beta} & \dfrac{\text{ch }\beta L + \cos\beta L}{2} \\[2mm]
\dfrac{\text{ch }\beta L - \cos\beta L}{2EI\beta^2} & \dfrac{\text{sh }\beta L + \sin\beta L}{2EI\beta} \\[2mm]
\dfrac{\text{sh }\beta L - \sin\beta L}{2EI\beta^3} & \dfrac{\text{ch }\beta L - \cos\beta L}{2EI\beta^2}
\end{bmatrix}
$$

$$
\begin{bmatrix}
\dfrac{EI\beta^2(\text{ch }\beta L - \cos\beta L)}{2} & \dfrac{EI\beta^3(\text{sh }\beta L + \sin\beta L)}{2} \\[2mm]
\dfrac{EI\beta(\text{sh }\beta L - \sin\beta L)}{2} & \dfrac{EI\beta^2(\text{ch }\beta L - \cos\beta L)}{2} \\[2mm]
\dfrac{\text{ch }\beta L + \cos\beta L}{2} & \dfrac{\beta(\text{sh }\beta L - \sin\beta L)}{2} \\[2mm]
\dfrac{\text{sh }\beta L + \sin\beta L}{2\beta} & \dfrac{\text{ch }\beta L + \cos\beta L}{2}
\end{bmatrix}
\begin{bmatrix} Q_0 \\ M_0 \\ \psi_0 \\ V_0 \end{bmatrix}
$$

where $\text{ch} = \cosh$, $\text{sh} = \sinh$.

Exercise 11: *Using the exact transfer matrix for a beam, rederive the relation for the frequencies of a clamped–free beam.*

The boundary conditions require

$$
Q_L = \frac{\cosh\beta L + \cos\beta L}{2} Q_0 + \beta\frac{\sinh\beta L - \sin\beta L}{2} M_0 = 0
$$

$$
M_L = \frac{\sinh\beta L + \sin\beta L}{2\beta} Q_0 + \frac{\cosh\beta L + \cos\beta L}{2} M_0 = 0
$$

Hence

$$
(\cosh\beta L + \cos\beta L)(\cosh\beta L + \cos\beta L)
$$
$$
- (\sinh\beta L + \sin\beta L)(\sinh\beta L - \sin\beta L) = 0
$$

The expansion of this expression gives

$$
1 + \cos\beta L \cosh\beta L = 0
$$

Exercise 12: *Consider the bar in longitudinal motion, see Figure 14, subjected* *to a step force which is constant at F for $t \leq 0$ and equal to zero for $t > 0$. Find $u(x, t)$ and $P(x, t)$.*

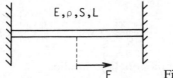

E,ρ,S,L

F   Figure 14

Since the bar is fixed at both ends,

$$u(x, t) = \sum_{n=1,2,\ldots} (A_n \cos \omega_n t + B_n \sin \omega_n t) \sin \frac{n\pi x}{L}$$

with

$$\omega_n = \frac{n\pi}{L} \sqrt{\frac{E}{\rho}}$$

The initial velocity distribution is given by

$$\frac{\partial u(x, t)}{\partial t}\bigg|_{t=0} = 0$$

Since

$$\frac{\partial u(x, t)}{\partial t} = \sum_{n=1,2,\ldots} (-\omega_n A_n \sin \omega_n t + \omega_n B_n \cos \omega_n t) \sin \frac{n\pi x}{L}$$

then at $t = 0$

$$\sum_{n=1,2,\ldots} \omega_n B_n \sin \frac{n\pi x}{L} = 0$$

Due to the orthogonality of the modes,

$$\int_0^L \omega_n B_n \sin^2 \frac{n\pi x}{L} dx = \frac{L\omega_n B_n}{2} = 0$$

hence $B_n = 0$.

The value of A_n is found from

$$u(x, t) = \sum_{n=1,2,\ldots} A_n \cos \omega_n t \sin \frac{n\pi x}{L}$$

which at $t = 0$ becomes

$$u(x, 0) = \sum_{n=1,2,\ldots} A_n \sin \frac{n\pi x}{L}$$

At the initial instant $t = 0$:

$$u(x, 0) = \frac{Fx}{2ES} \qquad \text{for} \qquad 0 < x < \frac{L}{2}$$

$$u(x, 0) = \frac{F(L - x)}{2ES} \qquad \text{for} \qquad \frac{L}{2} < x < L$$

which, taking account of orthogonality, implies

$$\frac{F}{2ES} \left[\int_0^{L/2} x \sin \frac{n\pi x}{L} \, dx + \int_{L/2}^L (L - x) \sin \frac{n\pi x}{L} \, dx \right] = A_n \int_0^L \sin^2 \frac{n\pi x}{L} \, dx$$

$$= \frac{A_n L}{2}$$

Since

$$\int_{L/2}^L (L - x) \sin \frac{n\pi x}{L} \, dx = \int_0^{L/2} x \sin \left(n\pi - \frac{n\pi x}{L} \right) dx$$

then

$$A_n = \frac{F}{ESL} \int_0^{L/2} x(1 - \cos n\pi) \sin \frac{n\pi x}{L} \, dx$$

Completing the calculations gives

$$A_n = \frac{2FL}{n^2 \pi^2 ES} (-1)^{(n-1)/2} \qquad \text{with } n = 1, 3, \ldots$$

$$u(x, t) = \frac{2FL}{\pi^2 ES} \sum_{n=1,3,\ldots} (-1)^{(n-1)/2} \frac{1}{n^2} \cos \omega_n t \sin \frac{n\pi x}{L}$$

and

$$P(x, t) = ES \frac{\partial u(x, t)}{\partial x}$$

$$= \frac{2F}{\pi} \sum_{n=1,3,\ldots} (-1)^{(n-1)/2} \frac{1}{n} \cos \omega_n t \cos \frac{n\pi x}{L}$$

The series representing $u(x, t)$ converges like $1/n^2$ and that for $P(x, t)$ like $1/n$. It is necessary to use more terms in calculating P than in calculating u. This observation has general applicability.

Exercise 13: *Beam on elastic foundation. A beam rests on an elastic foundation whose stiffness per unit length is given by $k = K/L$. Find the equation of motion, the general expression for the frequencies, and the specific expression for the frequencies of an S–S beam; see Figure 15.*

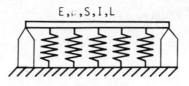

Figure 15

The first equation of (17) becomes

$$\rho S \frac{\partial^2 v}{\partial t^2} = -\frac{\partial Q}{\partial x} + Q_{ex} - kv$$

the others are unchanged. By elimination among these equations,

$$\frac{\partial^2}{\partial x^2}\left(EI \frac{\partial^2 v}{\partial x^2}\right) + \rho S \frac{\partial^2 v}{\partial t^2} + kv - Q_{ex} = 0$$

For free vibration and a constant cross-section, this becomes

$$EI \frac{\partial^4 v}{\partial x^4} + \rho S \frac{\partial^2 v}{\partial t^2} + kv = 0$$

Solutions are sought in the form

$$v(x, t) = \phi(x) \cdot f(t)$$

which gives

$$\frac{d^2 f(t)}{dt^2} + \omega^2 f(t) = 0$$

and

$$\frac{d^4 \phi(x)}{dx^4} + \frac{k - \omega^2 \rho S}{EI} \phi(x) = 0$$

Setting

$$\beta^* = \sqrt[4]{\frac{\rho S \omega^2 - k}{EI}}$$

one gets the same general solution as in the beam; namely:

$$\phi(x) = C \sin\beta^* x + D \cos \beta^* x + E \sinh \beta^* x + F \cosh \beta^* x$$

Using the boundary conditions, one obtains, as in section 4.2,

$$\beta_n^* L = X_n$$

from which

$$\omega_n = \frac{1}{L^2} \sqrt{\frac{EI}{\rho S}} \cdot \sqrt{X_n^4 + \frac{KL^3}{EI}}$$

For the S–S case, ϕ_n reduces to

$$\phi_n(x) = C_n \sin \beta_n^* x$$

and

$$\beta_n^* L = n\pi$$

so that

$$\omega_n = \frac{1}{L^2} \sqrt{\frac{EI}{\rho S}} \sqrt{(n\pi)^4 + \frac{KL^3}{EI}}$$

Exercise 14: *Consider the same system as in exercise 1. Use the Rayleigh–Ritz method to calculate the first two frequencies and their corresponding modes. Use a displacement function containing the terms* (x/L), $(x/L)^2$, $(x/L)^3$, $(x/L)^4$, *and program no. 5 of Chapter 7.*

The fixed–free bar has been previously analyzed with the same displacement functions; see (84). The spring gives the following supplementary stiffness matrix:

$$\frac{ES}{L} \begin{bmatrix} 1 & 1 & 1 & 1 \\ 1 & 1 & 1 & 1 \\ 1 & 1 & 1 & 1 \\ 1 & 1 & 1 & 1 \end{bmatrix}$$

Use of the computer program on the modified equation of motion gives

$$\omega_1 = \frac{2.029}{L} \sqrt{\frac{E}{\rho}}$$

$$\phi_1(x) = \frac{x}{L} + 0.06346 \left(\frac{x}{L}\right)^2 - 0.8993 \left(\frac{x}{L}\right)^3 + 0.2805 \left(\frac{x}{L}\right)^4$$

$$\omega_2 = \frac{4.932}{L} \sqrt{\frac{E}{\rho}}$$

$$\phi_2(x) = \frac{x}{L} - 0.9260 \left(\frac{x}{L}\right)^2 - 2.231 \left(\frac{x}{L}\right)^3 + 1.978 \left(\frac{x}{L}\right)^4$$

Exercise 15: *Consider the same system as in exercise 1. Use the Rayleigh–Ritz method to calculate the first two frequencies and their corresponding modes. Use a displacement function containing the terms* $\sin(\pi x/2L)$ *and* $\sin(3\pi x/2L)$.

The displacement is

$$u(x, t) = \sin \frac{\pi x}{2L} p_1 + \sin \frac{3\pi x}{2L} p_2$$

The kinetic energy is

$$T = \frac{1}{2} \int_0^L \rho S \left(\sin \frac{\pi x}{2L} \, p_1{}^\circ + \sin \frac{3\pi x}{2L} \, p_2{}^\circ \right)^2 dx$$

$$= \frac{\rho SL}{4} p_1{}^{\circ 2} + \frac{\rho SL}{4} p_2{}^{\circ 2}$$

The strain energy is

$$U = U_{\text{beam}} + U_{\text{spring}}$$

$$U = \frac{1}{2} \int_0^L ES \left(\frac{\pi}{2L} \cos \frac{\pi x}{2L} p_1 + \frac{3\pi}{2L} \cos \frac{3\pi x}{2L} p_2 \right)^2 dx + \frac{ES}{2L} (p_1 - p_2)^2$$

$$= \frac{ES}{L} \left(\frac{\pi^2}{16} + \frac{1}{2} \right) p_1^2 + \frac{ES}{L} \left(\frac{9\pi^2}{16} + \frac{1}{2} \right) p_2^2 - \frac{ES}{L} p_1 p_2$$

The application of Lagrange's equations gives

$$\rho SL \begin{bmatrix} \dfrac{1}{2} & 0 \\ 0 & \dfrac{1}{2} \end{bmatrix} \begin{bmatrix} p_1{}^{\infty} \\ p_2{}^{\infty} \end{bmatrix} + \frac{ES}{L} \begin{bmatrix} \dfrac{\pi^2}{8} + 1 & -1 \\ -1 & \dfrac{9\pi^2}{8} + 1 \end{bmatrix} \begin{bmatrix} p_1 \\ p_2 \end{bmatrix} = 0$$

from which the frequencies and modes are

$$\omega_1 = \frac{2.066}{L} \sqrt{\frac{E}{\rho}} \, ; \qquad \phi_1(x) = \sin \frac{\pi x}{2L} + 0.1003 \sin \frac{3\pi x}{2L}$$

$$\omega_2 = \frac{4.940}{L} \sqrt{\frac{E}{\rho}} \, ; \qquad \phi_2(x) = \sin \frac{\pi x}{2L} - 9.970 \sin \frac{3\pi x}{2L}$$

Exercise 16: *Consider a fixed–free bar in longitudinal motion which has a concentrated tip mass of $m = \rho SL/10$; see Figure 16. Use the Rayleigh–Ritz method and the same displacement function as in exercise 15 to calculate the frequencies and modes.*

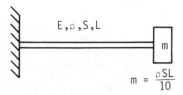

$$E, \rho, S, L$$

$$m$$

$$m = \frac{\rho SL}{10}$$

Figure 16

The strain energy is

$$U = \frac{ES}{2} \left[\frac{\pi^2}{8L} p_1^2 + \frac{9\pi^2}{8L} p_2^2 \right]$$

and the kinetic energy is

$$T = T_{beam} + T_{mass}$$

$$= \frac{\rho SL}{4}(p_1{}^{o2} + p_2{}^{o2}) + \frac{\rho SL}{20}(p_1{}^{o2} + p_2{}^{o2} - 2p_1{}^{o}p_2{}^{o})$$

Applying Lagrange's equations gives

$$\rho SL \begin{bmatrix} 0.6 & -0.1 \\ -0.1 & 0.6 \end{bmatrix} \begin{bmatrix} p_1{}^{\infty} \\ p_2{}^{\infty} \end{bmatrix} + \frac{ES}{L} \begin{bmatrix} 1.234 & 0 \\ 0 & 11.10 \end{bmatrix} \begin{bmatrix} p_1 \\ p_2 \end{bmatrix} = 0$$

from which

$$\omega_1 = \frac{1.431}{L}\sqrt{\frac{E}{\rho}} \qquad \text{with the mode} \qquad \begin{bmatrix} 1 \\ -0.0208 \end{bmatrix}$$

Hence

$$\phi_1(x) = \sin\frac{\pi x}{2L} - 0.0208 \sin\frac{3\pi x}{2L}$$

and also

$$\omega_2 = \frac{4.370}{L}\sqrt{\frac{E}{\rho}} \qquad \text{with the mode} \qquad \begin{bmatrix} 1 \\ 5.354 \end{bmatrix}$$

Hence

$$\phi_2(x) = \sin\frac{\pi x}{2L} + 5.354 \sin\frac{3\pi x}{2L}$$

Exercise 17: *Use Rayleigh's method to calculate the frequency for the beam shown in Figure* 17. *Use the displacement function* $3(x/L)^2 - (x/L)^3$.

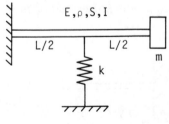

$$E, \rho, S, I$$

$$L/2 \qquad L/2$$

$$m$$

$$k$$

Figure 17

As suggested,

$$v(x, t) = \left[3\left(\frac{x}{L}\right)^2 - \left(\frac{x}{L}\right)^3\right]p$$

The kinetic and strain energy of the beam have been calculated in section 4.3. Then

$$T = 0.4714\rho SLp^{\circ 2} + 2mp^{\circ 2}$$

$$U = \frac{6EI}{L^3}p^2 + 0.1953kp^2$$

Then the equation of motion is

$$(4m + 0.9428\rho SL)p^{\infty} + \left(\frac{12EI}{L^3} + 0.3906k\right)p = 0$$

and

$$\omega = \sqrt{\frac{12EI/L^3 + 0.3906k}{4m + 0.9428\rho SL}}$$

Exercise 18: *Effect of axial force. A clamped–free beam is subjected to a constant axial force F_0. Use Rayleigh's method to calculate the frequency with the same displacement function as in exercise 17.*

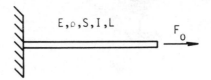

Figure 18

From (79) and (80),

$$T = 0.4714\rho SLp^{\circ 2}$$

$$U = \frac{6EI}{L^3}p^2$$

It is necessary to add the influence of axial force to the strain energy U. From (141), the supplementary strain energy is

$$U_{F_0} = \frac{F_0}{2}\int_0^L \left(\frac{\partial v}{\partial x}\right)^2 dx = \frac{F_0}{2}\int_0^L \left(\frac{6x}{L^2} - \frac{3x^2}{L^3}\right)^2 p^2 dx$$

$$= \frac{2.4F_0}{L}p^2$$

The system equation then becomes

$$0.9428\rho SLp^{\infty} + \left(\frac{12EI}{L^3} + \frac{4.8F_0}{L}\right)p = 0$$

and the frequency is

$$\omega = \sqrt{\frac{12EI/L^3 + 4.8F_0/L}{0.9428\rho SL}}$$

The numerator of ω vanishes for $F_0 = -2.5EI/L^2$. This gives a good approximate value of the first Euler buckling load, since the exact value is

$$-\frac{\pi^2}{4}\frac{EI}{L^2} = -2.46\frac{EI}{L^2}$$

The vibration mode becomes the buckling mode.

Exercise 19: *A beam simply supported at each end has a spring of stiffness* $k = 40EI/L^3$ *at its quarter-point; see Figure* 19. *Use Rayleigh's method to calculate the lowest frequency.*

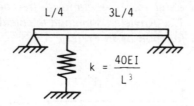

Figure 19

Since the beam is S–S, an appropriate displacement function is $\sin(\pi x/L)$. From the results of exercise 9:

$$T_{\text{beam}} = \frac{\rho SL}{4}p^{\circ 2}$$

$$U_{\text{beam}} = \frac{\pi^4}{4}\frac{EI}{L^3}p^2$$

The strain energy for the spring is

$$U_{\text{spring}} = \frac{20EI}{L^3}\left(\sin\frac{\pi}{4}\right)^2 p^2 = \frac{10EI}{L^3}p^2$$

The strain energy of the system is then

$$U = U_{\text{beam}} + U_{\text{spring}} = \frac{EI}{L^3}\left(\frac{\pi^4}{4} + 10\right)p^2$$

hence the frequency is

$$\omega = \frac{1}{L^2}\sqrt{40 + \pi^4}\sqrt{\frac{EI}{\rho S}} = \frac{11.72}{L^2}\sqrt{\frac{EI}{\rho S}}$$

The exact solution obtained from the transfer matrix program no. 10 is $11.62/L^2\sqrt{EI/\rho S}$.

Exercise 20: *Consider the same system as in exercise* 19. *Use the Rayleigh–Ritz method to calculate the first frequency and its corresponding mode. Use the displacement function* $v(x, t) = \sin(\pi x/L)p_1 + \sin(2\pi x/L)p_2$.

For the beam: see exercise 9, from which

$$T_{beam} = \frac{\rho SL}{4} p_1^{\circ 2} + \frac{\rho SL}{4} p_2^{\circ 2}$$

$$U_{beam} = \frac{\pi^4}{4} \frac{EI}{L^3} (p_1^2 + 16p_2^2)$$

For the spring:

$$U_{spring} = \frac{20EI}{L^3} \left[\frac{\sqrt{2}}{2} p_1 + p_2 \right]^2$$

$$= \frac{20EI}{L^3} \left(\frac{p_1^2}{2} + p_2^2 + \sqrt{2}\, p_1 p_2 \right)$$

Application of Lagrange' equations gives

$$\frac{\rho SL}{2} \begin{bmatrix} 1 & 0 \\ 0 & 1 \end{bmatrix} \begin{bmatrix} p_1^{\circ\circ} \\ p_2^{\circ\circ} \end{bmatrix} + \frac{EI}{L^3} \begin{bmatrix} \dfrac{\pi^4}{2} + 20 & 20\sqrt{2} \\ 20\sqrt{2} & 8\pi^4 + 40 \end{bmatrix} \begin{bmatrix} p_1 \\ p_2 \end{bmatrix} = 0$$

from which

$$\omega = \frac{11.63}{L^2} \sqrt{\frac{EI}{\rho S}} \quad \text{with the mode} \quad \begin{bmatrix} 1 \\ -0.0376 \end{bmatrix}$$

Hence

$$\phi(x) = \frac{\sin \pi x}{L} - 0.0376 \sin \frac{2\pi x}{L}$$

Note that this result for frequency is more accurate than that of exercise 19.

Exercise 21: *Consider the clamped–free beam shown in Figure* 20. *Use the Rayleigh–Ritz method to calculate the first two frequencies and modes. Use the displacement function:* $v(x, t) = (x/L)^2 p_1 + (x/L)^3 p_2$.

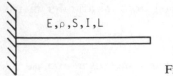

Figure 20

The kinetic and strain energies are respectively

$$T = \frac{1}{2} \int_0^L \rho S \left[\left(\frac{x}{L} \right)^2 p_1^{\circ} + \left(\frac{x}{L} \right)^3 p_2^{\circ} \right]^2 dx$$

$$= \frac{\rho SL}{10} p_1^{\circ 2} + \frac{\rho SL}{6} p_1^{\circ} p_2^{\circ} + \frac{\rho SL}{14} p_2^{\circ 2}$$

$$U = \frac{1}{2} \int_0^L EI \left(\frac{2}{L^2} p_1 + \frac{6x}{L^3} p_2 \right)^2 dx$$

$$= \frac{2EI}{L^3} p_1^2 + \frac{6EI}{L^3} p_1 p_2 + \frac{6EI}{L^3} p_2^2$$

Application of Lagrange's equations gives

$$\rho SL \begin{bmatrix} \frac{1}{5} & \frac{1}{6} \\ \frac{1}{6} & \frac{1}{7} \end{bmatrix} \begin{bmatrix} p_1^{\infty} \\ p_2^{\infty} \end{bmatrix} + \frac{EI}{L^3} \begin{bmatrix} 4 & 6 \\ 6 & 12 \end{bmatrix} \begin{bmatrix} p_1 \\ p_2 \end{bmatrix} = 0$$

from which

$$\omega_1 = \frac{3.533}{L^2} \sqrt{\frac{EI}{\rho S}} \qquad \text{with the mode} \qquad \begin{bmatrix} 1 \\ -0.3837 \end{bmatrix}$$

and

$$\omega_2 = \frac{34.81}{L^2} \sqrt{\frac{EI}{\rho S}} \qquad \text{with the mode} \qquad \begin{bmatrix} -0.8221 \\ 1 \end{bmatrix}$$

Hence

$$\phi_1(x) = \left(\frac{x}{L} \right)^2 - 0.3837 \left(\frac{x}{L} \right)^3$$

$$\phi_2(x) = -0.8221 \left(\frac{x}{L} \right)^2 + \left(\frac{x}{L} \right)^3$$

Exercise 22: *Consider a beam which is simply supported at its two ends and which has a concentrated mass of $m = \rho SL/5$ at its quarter-point; see Figure 21. Use the Rayleigh–Ritz method to calculate its first two frequencies and their corresponding modes. Use a displacement function containing the terms* $\sin(\pi x/L)$, $\sin(2\pi x/L)$, $\sin(3\pi x/L)$.

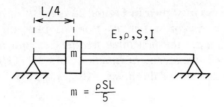

$$m = \frac{\rho SL}{5}$$

Figure 21

The results of exercise 9 give

$$U_{\text{beam}} = \frac{\pi^4 EI}{4L^3}(p_1^2 + 16p_2^2 + 81p_3^2)$$

$$T_{\text{beam}} = \frac{\rho SL}{4}(p_1^{\circ 2} + p_2^{\circ 2} + p_3^{\circ 2})$$

To the kinetic energy of the beam must be added the kinetic energy of the concentrated mass:

$$T_{\text{mass}} = \frac{\rho SL}{10}\left(\frac{\sqrt{2}}{2}p_1^\circ + p_2^\circ + \frac{\sqrt{2}}{2}p_3^\circ\right)^2$$

Using these energy expressions in Lagrange's equations gives

$$\rho SL\begin{bmatrix} 0.6 & 0.1414 & 0.1 \\ 0.1414 & 0.7 & 0.1414 \\ 0.1 & 0.1414 & 0.6 \end{bmatrix}\begin{bmatrix} p_1^{\circ\circ} \\ p_2^{\circ\circ} \\ p_3^{\circ\circ} \end{bmatrix} + \frac{\pi^4 EI}{2L^3}\begin{bmatrix} 1 & 0 & 0 \\ 0 & 16 & 0 \\ 0 & 0 & 81 \end{bmatrix}\begin{bmatrix} p_1 \\ p_2 \\ p_3 \end{bmatrix} = 0$$

from which

$$\omega_1 = \frac{8.991}{L^2}\sqrt{\frac{EI}{\rho S}} \qquad \text{with the mode} \qquad \begin{bmatrix} 1 \\ 0.0159 \\ 0.0021 \end{bmatrix}$$

and

$$\omega_2 = \frac{34.13}{L^2}\sqrt{\frac{EI}{\rho S}} \qquad \text{with the mode} \qquad \begin{bmatrix} 1 \\ -3.835 \\ -0.1587 \end{bmatrix}$$

Hence

$$\phi_1(x) = \sin\frac{\pi x}{L} + 0.0159\sin\frac{2\pi x}{L} + 0.0021\sin\frac{3\pi x}{L}$$

$$\phi_2(x) = \sin\frac{\pi x}{L} - 3.835\sin\frac{2\pi x}{L} - 0.1587\sin\frac{3\pi x}{L}$$

Exercise 23: *The system shown in Figure 22 consists of a clamped–free beam with a spring–mass system suspended from its free end. Use the Rayleigh–Ritz method to calculate the first three frequencies and modes of this system. For the beam use the displacement function:* $v(x, t) = (x/L)^2 p_1 + (x/L)^3 p_2$ *and let p_3 be the displacement of the mass. Also, $k = EI/L^3$ and $m = \rho SL/7$.*

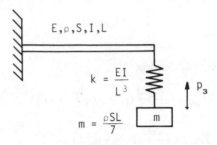

Figure 22

The kinetic energy of the mass m and the strain energy of the spring k must be added to the energies expressions found in exercise 21. These additions are

$$U_{\text{spring}} = \tfrac{1}{2}k(p_1 + p_2 - p_3)^2$$

$$= \frac{EI}{2L^3}(p_1^2 + p_2^2 + p_3^2 + 2p_1p_2 - 2p_2p_3 - 2p_1p_3)$$

$$T_{\text{mass}} = \tfrac{1}{2}mp_3^{\circ 2} = \frac{\rho SL}{14}p_3^{\circ 2}$$

Application of Lagrange's equations gives

$$\rho SL \begin{bmatrix} \dfrac{1}{5} & \dfrac{1}{6} & 0 \\[6pt] \dfrac{1}{6} & \dfrac{1}{7} & 0 \\[6pt] 0 & 0 & \dfrac{1}{7} \end{bmatrix} \begin{bmatrix} p_1^{\circ\circ} \\ p_2^{\circ\circ} \\ p_3^{\circ\circ} \end{bmatrix} + \frac{EI}{L^3} \begin{bmatrix} 5 & 7 & -1 \\ 7 & 13 & -1 \\ -1 & -1 & 1 \end{bmatrix} \begin{bmatrix} p_1 \\ p_2 \\ p_3 \end{bmatrix} = 0$$

from which

$$\omega_1 = \frac{2.143}{L^2}\sqrt{\frac{EI}{\rho S}}; \qquad \phi_1 = \begin{bmatrix} 0.5308 \\ -0.1871 \\ 1 \end{bmatrix}$$

$$\omega_2 = \frac{4.346}{L^2}\sqrt{\frac{EI}{\rho S}}; \qquad \phi_2 = \begin{bmatrix} 1 \\ -0.4077 \\ -0.3486 \end{bmatrix}$$

$$\omega_3 = \frac{34.92}{L^2} \sqrt{\frac{EI}{\rho S}} ; \qquad \phi_3 = \begin{bmatrix} -0.8215 \\ 1 \\ -0.0010 \end{bmatrix}$$

The exact frequencies calculated from program no. 10 are

$$\omega_1 = \frac{2.143}{L^2} \sqrt{\frac{EI}{\rho S}} ; \qquad \omega_2 = \frac{4.320}{L^2} \sqrt{\frac{EI}{\rho S}} ; \qquad \omega_3 = \frac{22.13}{L^2} \sqrt{\frac{EI}{\rho S}}$$

Exercise 24: *A clamped–free bar is subjected to a longitudinal sinusoidal exciting force F sin Ωt at its free end as shown in Figure 23. Find the analytic solution for steady-state motion.*

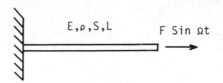

Figure 23

The solution to the equation of motion is

$$u(x, t) = (A \sin \Omega t + B \cos \Omega t)\left(C \sin \Omega \sqrt{\frac{\rho}{E}} x + D \cos \Omega \sqrt{\frac{\rho}{E}} x\right)$$

since $u(0, t) = 0$, $D = 0$ and

$$u(x, t) = (A^* \sin \Omega t + B^* \cos \Omega t) \sin \Omega \sqrt{\frac{\rho}{E}} x$$

At $x = L$,

$$P(L, t) = ES \frac{\partial u}{\partial x}\bigg|_{x=L} = F \sin \Omega t$$

thus

$$ES\Omega \sqrt{\frac{\rho}{E}} \cos \Omega \sqrt{\frac{\rho}{E}} L (A^* \sin \Omega t + B^* \cos \Omega t) = F \sin \Omega t$$

which requires

$$B^* = 0, \qquad \text{and} \qquad A^* = \frac{F}{ES\Omega \sqrt{\frac{\rho}{E}} \cos \Omega \sqrt{\frac{\rho}{E}} L}$$

Finally,

$$u(x, t) = \frac{F}{ES\Omega \sqrt{\frac{\rho}{E}} \cos \Omega \sqrt{\frac{\rho}{E}} L} \sin \Omega \sqrt{\frac{\rho}{E}} x \sin \Omega t$$

Exercise 25: *A beam simply supported at each end is subjected to a force F sin Ωt at its midpoint, as in Figure 24. Using the results of exercise 9, find*

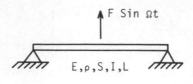

Figure 24

the steady-state motion. Plot

$$\left|\frac{V(L/2)}{F}\right|\frac{EI}{L^3} \qquad \textit{as a function of} \qquad f^* = \frac{f}{\frac{1}{L^2}\sqrt{\frac{EI}{\rho S}}}.$$

From exercise 9:

$$T = \sum_{n=1}^{\infty} T_n = \frac{\rho SL}{4}\sum_{n=1}^{\infty} p_n^{\circ 2} = \frac{\rho SL}{4}(p_1^{\circ 2} + p_2^{\circ 2} + \ldots)$$

and

$$U = \sum_{n=1}^{\infty} U_n = \frac{\pi^4 EI}{4L^3}\sum_{n=1}^{\infty} n^4 p_n^2 = \frac{\pi^4 EI}{4L^3}(p_1^2 + 16p_2^2 + \ldots)$$

The generalized forces are deduced from the virtual work of the external force $F \sin \Omega t$.

$$\delta W = F \sin \Omega t \sum_{n=1}^{\infty} \sin\frac{n\pi}{2}\,\delta p_n = F \sin \Omega t \sum_{1,3,5,\ldots} (-1)^{(n-1)/2}\,\delta p_n$$

Application of Lagrange's equations gives a set of uncoupled equations of motion:

$$\frac{\rho SL}{2}\,p_n{}^{\circ\circ} + \frac{\pi^4 EIn^4}{2L^3}\,p_n = (-1)^{(n-1)/2}F \sin \Omega t$$

for $n = 1, 3, 5, \ldots$.

The equations of motion for n even do not contribute to the steady-state solutions since the generalized forces are zero. The steady-state solution is

$$p_n = \frac{2L^3}{\pi^4 EIn^4}\cdot\frac{(-1)^{(n-1)/2}F \sin \Omega t}{1-(\Omega/\omega_n)^2}$$

Hence

$$v(x, t) = \sum_{n=1,3,\ldots} \phi_n(x)p_n = \sum_{n=1,3,\ldots} \sin\frac{n\pi x}{L}\,p_n = V(x)\sin \Omega t$$

The quantities M and $Q = \partial M/\partial x$ correspond to the second and third derivatives of the variable $v(x, t)$ with respect to space, x. It is then necessary to sum more modes to achieve accurate values of M and Q than it is for the lateral deflection v. The results are plotted in Figure 25.

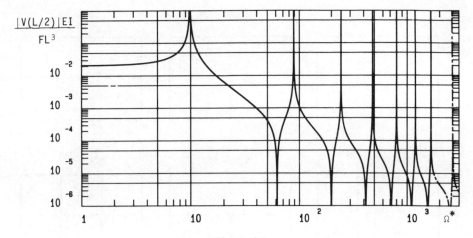

Figure 25

Exercise 26: *For the case of a rectangular plate of sides a, b which is simply supported on all four edges, calculate ω_{rs} using the displacement function given in equation* (112). *Plot the nodal lines of the first six modes for the case* $b = 2a/3$.

Using the expressions (113), (114) in Lagrange's equations gives

$$\omega_{rs} = \pi^2 \sqrt{\frac{D}{\rho h}} \left(\frac{r^2}{a^2} + \frac{s^2}{b^2} \right)$$

For $b = 2a/3$:

$$\omega_{rs} = \frac{\pi^2 h}{a^2} \sqrt{\frac{E}{12(1-\nu^2)\rho}} \cdot \left(r^2 + \frac{9s^2}{4} \right)$$

$$= A\omega^*_{rs}$$

where

$$A = \frac{\pi^2 h}{a^2} \sqrt{\frac{E}{12(1-\nu^2)\rho}} \quad \text{and} \quad \omega^*_{rs} = r^2 + \frac{9s^2}{4}$$

The nodal lines are plotted in Figure 26.

Exercise 27: *Consider the case of a circular plate of radius R and constant thickness h which is clamped at its center and free at its circumference. Use Rayleigh's method and the displacement function:*

$$w(r, \theta, t) = p_n [3(r/R)^2 - (r/R)^3] \cos n\theta$$

to calculate the frequencies f_n.

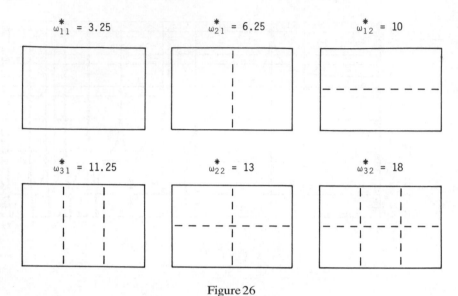

Figure 26

Substituting $w(r, \theta, t)$ in the expressions for strain energies (104) and kinetic energy (105) and completing all the calculations gives, for $n \neq 0$:

$$T_n = 1.206 \rho h R p_n^{\circ 2}$$

$$U_n = 0.326 \frac{D}{R^2} [11n^4 - n^2(26 + 32\nu) + (45 + 36\nu)] p_n^2$$

The frequencies are then

$$f_n = \frac{0.026}{\sqrt{1 - \nu^2}} \frac{h}{R^2} \sqrt{\frac{E}{\rho}} \sqrt{11n^4 - n^2(26 + 32\nu) + (45 + 36\nu)}$$

The formula for frequency is valid for $n = 0$ but in this case the kinetic and strain energies have their coefficients multiplied by a factor of two.

Exercise 28: The system shown in Figure 27 represents a rotor simply supported at its ends on rigid bearings. The disk is rigid with properties $I_{Dx} = 0.1225$ kg·m² (13 slug·in²), $I_{Dy} = 0.2450$ kg·m² (26 slug·in²), $M_D = 7.85$ kg (0.538 slug). The shaft has a length $L = 0.4$ m, (15.75 in) and area moment of inertia of its cross-section is: $I_x = I_z = 0.49 \times 10^{-9}$ m⁴ (1.177 × 10⁻³ in⁴). Young's modulus is $E = 2 \times 10^{11}$ N/m² (29 × 10⁶ psi). The mass of the shaft is neglected. Taking the exact modes of a simply supported beam as displacement functions use the Rayleigh–Ritz method to calculate the first two frequencies of the rotor as a function of the angular velocity of rotation $\Omega_r = 2\pi N$ where N is the shaft speed in rev/sec.

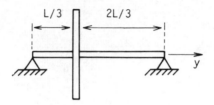

Figure 27

The displacement functions are

$$u(y, t) = \sin \frac{\pi y}{L} p_1$$

$$w(y, t) = \sin \frac{\pi y}{L} p_2$$

where

$$\theta = \frac{\partial w}{\partial y} = \frac{\pi}{L} \cos \frac{\pi y}{L} p_2$$

$$\psi = -\frac{\partial u}{\partial y} = -\frac{\pi}{L} \cos \frac{\pi y}{L} p_1$$

Using equations (131) the kinetic energy of the disk in SI units is

$$T_D = 3.888(p_1^{\circ 2} + p_2^{\circ 2}) - 3.778\Omega_r p_1^{\circ} p_2 + 0.1225\Omega_r^2$$

Using equation (145), the strain energy of the shaft in SI units is

$$U_s = 3.729 \times 10^4 (p_1^2 + p_2^2)$$

Applying Lagrange's equations gives

$$7.777 p_1^{\circ\circ} - 3.778\Omega_r p_2^{\circ} + 7.458 \times 10^4 p_1 = 0$$
$$7.777 p_2^{\circ\circ} + 3.778\Omega_r p_1^{\circ} + 7.458 \times 10^4 p_2 = 0$$

Solutions are sought in the form $p_1 = P_1 e^{rt}$ and $p_2 = P_2 e^{rt}$ which gives the characteristic equation

$$r^4 + (1.918 \times 10^4 + 0.2360\Omega_r) r^2 + 9.197 \times 10^7 = 0$$

Using $r = \pm j\omega$ in this gives $\omega = \omega(\Omega_r)$ and the frequency $f(\Omega_r) = \omega(\Omega_r)/2\pi$ (see Figure 28) for curves A and B which are the desired solution. Points (1) and (2) correspond to the equality between the speed of rotation of the rotor and its frequency. They are 12.8 Hz and 21.7 Hz. The results shown in the figure are the same in SI and English units.

Exercise 29: *Critical speed due to unbalance. The data are the same as in exercise 28. The unblance mass is $m \ll M_D$. This mass situated on z (Figure*

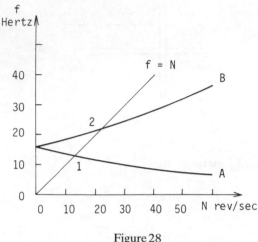

Figure 28

8), *is located on the disk at a radial distance d from its geometric center. Calculate the kinetic energy of this mass, find the equation of motion of the rotor system and its steady-state response.*

In the fixed coordinate system, the coordinates of the unbalanced mass are

$$\sin\frac{\pi}{3}p_1 + d\sin\Omega_r t = 0.8660p_1 + d\sin\Omega_r t$$

$$\sin\frac{\pi}{3}p_2 + d\cos\Omega_r t = 0.8660p_2 + d\cos\Omega_r t$$

The kinetic energy of the unbalance mass is

$$T_m = \frac{m}{2}(0.75p_1^{\circ 2} + 0.75p_2^{\circ 2} + 1.732d\Omega_r\cos\Omega_r t p_1^{\circ}$$

$$- 1.732d\Omega_r\sin\Omega_r t p_2^{\circ} + d^2\Omega_r^2)$$

Hence

$$\frac{d}{dt}\left(\frac{\partial T_m}{\partial p_1^{\circ}}\right) - \frac{\partial T_m}{\partial p_1} = 0.75\,mp_1^{\circ\circ} - 0.8660d\Omega_r^2\sin\Omega_r t$$

$$\frac{d}{dt}\left(\frac{\partial T_m}{\partial p_2^{\circ}}\right) - \frac{\partial T_m}{\partial p_2} = 0.75mp_2^{\circ\circ} - 0.8660d\Omega_r^2\cos\Omega_r t$$

which are to be added to the equations of the motion found in exercise 28. However, since $m \ll M_D$ the terms of $0.75mp_1^{\circ\circ}$ and $0.75mp_2^{\circ\circ}$ can be neglected. The equations of motion for the rotor system with a mass unbalance of m are then, in SI units:

$$7.777p_1^{\infty} - 3.778\Omega_r p_2^{\circ} + 7.458 \times 10^4 p_1 = 0.8660 md\Omega_r^2 \sin \Omega_r t$$
$$7.777p_2^{\infty} + 3.778\Omega_r p_1^{\circ} + 7.458 \times 10^4 p_2 = 0.8660 md\Omega_r^2 \cos \Omega_r t$$

If $\alpha = 0.1114 md\Omega_r^2$, these can be written

$$p_1^{\infty} - 0.4858\Omega_r p_2^{\circ} + 9.590 \times 10^3 p_1 = \alpha \sin \Omega_r t$$
$$p_2^{\infty} + 0.4858\Omega_r p_1^{\circ} + 9.590 \times 10^3 p_2 = \alpha \cos \Omega_r t$$

For steady-state response,

$$p_1 = P_1 \sin \Omega_r t$$
$$p_2 = P_2 \cos \Omega_r t$$

which after some algebra yields

$$P_1 = P_2 = \frac{\alpha}{-0.5142\Omega_r^2 + 9.590 \times 10^3}$$

The critical speed is the value of Ω_r for which the steady-state response grows without bound. This corresponds to the value of Ω_r which makes the denominator of the above expression vanish, i.e. $\Omega_r = 136.6$ rad/sec or $N = 136.3/2\pi = 21.7$ rev/sec $= 21.7$ Hz. Note that the frequency 12.8 Hz found in exercise 28 is not a critical speed.

Exercise 30: *Consider the system of exercise 28 with a constant axial force F_0 applied to the shaft. Using the same method as in exercise 28, find the equations of motion of the system and plot the dependence of the first two frequencies on N for the values $F_0 = 10^4$ newtons (2248 lbf).*

One has

$$\frac{\partial w}{\partial y} = \frac{\pi}{L} \cos \frac{\pi y}{L} p_2; \qquad \frac{\partial u}{\partial y} = \frac{\pi}{L} \cos \frac{\pi y}{L} p_1$$

hence from equation (145) the additional strain energy due to the axial force is

$$U = \frac{F_0 \pi^2}{4L} (p_1^2 + p_2^2)$$

$$= 2.4167 \frac{F_0}{L} (p_1^2 + p_2^2)$$

The equations of motion of the system are then modified to:

$$7.777p_1^{\infty} - 3.778p_2^{\circ}\Omega_r + (12.34F_0 + 7.458 \times 10^4)p_1 = 0$$
$$7.777p_2^{\infty} + 3.778p_1^{\circ}\Omega_r + (12.34F_0 + 7.458 \times 10^4)p_2 = 0$$

in SI units. The frequencies are found as in exercise 28 and are plotted in Figure 29.

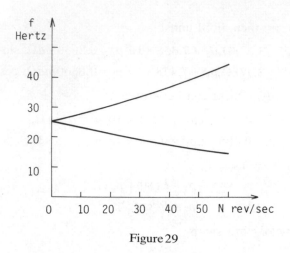

Figure 29

Exercise 31: *Consider a clamped–free beam whose deflection is given by equation (78). Use program no. 12 of Chapter 7 to calculate the kinetic and strain energies for the cases of 2, 3, and 4 Gauss points.*

The results of the computation are:

n	2	3	4
$\dfrac{L^3 U}{EIp^2}$	6	6	6
$\dfrac{T}{\rho S L p^{\circ 2}}$	0.4769	0.4713	0.4714

In using n Gauss points, one gets an exact value of the integral of a polynomial of degree $2n-1$. U and T correspond, respectively, to the integration of a polynomial of degree 2 and degree 6. Therefore, it is sufficient to take $n=2$ to calculate U, and $n=4$ to calculate T.

Exercise 32: *Consider a clamped–free beam having the displacement function $v(x, t) = (1 - \cos \pi x/2L)p$. Calculate the kinetic energy, the strain energy, and the frequency ω using program no. 12 with 4 Gauss points.*

The solution is

$$T = 0.1134 \rho S L p^{\circ 2}$$

$$U = 1.522 \frac{EI}{L^3} p^2$$

$$\omega = \frac{3.664}{L^2} \sqrt{\frac{EI}{\rho S}}$$

Exercise 33: *Discretization. Consider the same system as in exercise 1. The bar is modeled by 5 elements of length $L/5$. Each element is modeled by a mass ($\rho SL/10$)–spring ($5ES/L$)–mass ($\rho SL/10$) system. Write the overall system mass and stiffness matrices. Using program no. 5 calculate the first two frequencies and modes.*

$$
M = \frac{\rho SL}{5}
\begin{bmatrix}
1 & & & & \text{zero} \\
& 1 & & & \\
& & 1 & & \\
& & & 1 & \\
\text{zero} & & & & \frac{1}{2}
\end{bmatrix}
\qquad
K = \frac{5ES}{L}
\begin{bmatrix}
2 & -1 & & & \text{Zero} \\
-1 & 2 & -1 & & \\
& -1 & 2 & -1 & \\
& & -1 & 2 & -1 \\
\text{zero} & & & -1 & \frac{6}{5}
\end{bmatrix}
$$

$$
\omega_1 = \frac{2.024}{L}\sqrt{\frac{E}{\rho}}, \qquad \phi_1 = [1, 1.836, 2.371, 2.518, 2.252]^t
$$

$$
\omega_2 = \frac{4.748}{L}\sqrt{\frac{E}{\rho}}, \qquad \phi_2 = [1, 1.098, 0.2062, -0.8718, -1.164]^t
$$

5

Calculation Using Finite Elements

In most cases, the complex geometries and nonclassical boundary conditions of real structures require that numerical methods must be used to determine their static or dynamic behavior. For such cases, the finite element method is widely used and some of its aspects are presented in this chapter.

5.1 Derivation of Equations

The finite element method can be presented simply in terms of the following steps:

(a) the structure is divided into elements of finite size, called finite elements, which are connected at certain points, called nodal points or nodes, situated on the boundary of the element;

(b) after making a reasonable hypothesis for the displacement field of element i, the kinetic energy T_i, the strain energy U_i, and the dissipation function R_i are calculated for that element as a function of the nodal point displacements

(c) if the structure is composed of N elements, then:

$$T = \sum_{i=1}^{N} T_i$$

$$U = \sum_{i=1}^{N} U_i \qquad (1)$$

$$R = \sum_{i=1}^{N} R_i$$

The generalized forces are determined by writing the virtual work of the external forces.

The application of Lagrange's equations then allows the differential equations of motion of the whole structure to be obtained. In order to simplify the presentation of the method in what follows, it will be assumed that there is no dissipation of energy, i.e. $R_i = 0$ for all i.

5.1.1   Discretization: the division of the structure into finite elements

First, the type and distribution of finite elements must be chosen. This choice must take into account the geometry and the behavior of the structure. An effective and efficient discretization of a structure requires a great deal of experience. In particular, the analyst must take into account geometric discontinuities, material discontinuities, boundary conditions, and the forces applied to the structure.

For the accurate calculations of frequencies and modes or the calculation of dynamic response, the distribution of the finite elements, the so-called mesh, can be relatively coarse and regular. The mesh must be finer to calculate the stresses accurately and must be transformed into an even finer mesh in regions of stress concentation.

5.1.2   Strain energy – The stiffness matrix

In order to simplify the notation, the index i which designates quantities associated with the ith element is omitted in the following derivation.

The general expression for strain energy of an element is

$$U = \frac{1}{2} \int_\tau \varepsilon^t \sigma \, d\tau \tag{2}$$

The displacement vector d of an arbitrary point of the element is related to the nodal displacement vector δ of the element by means of a matrix N. The matrix N is generated from the hypothesis about the displacement field inside the element. The relation between d and δ is

$$d = N\delta \tag{3}$$

By differentiating (3), the relation between strains within the element and the nodal displacements is obtained:

$$\varepsilon = B\delta \tag{4}$$

In the case where there are no initial stresses, the stresses and strains are related by

$$\sigma = D\varepsilon \tag{5}$$

where D is a square symmetric matrix whose elements depend on the mechanical characteristics of the material, usually Young's modulus E and

Poisson's ratio ν. Using (5) and (4), equation (2) becomes

$$U = \frac{1}{2} \int_\tau (B\delta)^t DB\delta \, d\tau \tag{6}$$

$$= \tfrac{1}{2}\delta^t \left[\int_\tau B^t DB \, d\tau \right] \delta \tag{7}$$

which can be written

$$U = \tfrac{1}{2}\delta^t k\delta \tag{8}$$

with

$$k = \int_\tau B^t DB \, d\tau \tag{9}$$

being the stiffness matrix of the element. Matrix k is symmetric because if D is symmetric, the matrix product $B^t DB$ is also symmetric.

The determination of the strain energy U and stiffness matrix k will be illustrated for the cases of a bar in longitudinal motion and a beam in bending.

Bar in longitudinal motion

An element of the bar is shown in Figure 1. It has two nodes with one degree of freedom at each node; namely u, the longitudinal displacement.

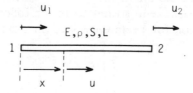

Figure 1   A bar in longitudinal motion

The nodal displacement vector is

$$\delta = \begin{bmatrix} u_1 \\ u_2 \end{bmatrix} \tag{10}$$

Since there are two nodal displacements, the displacement hypothesis is chosen to have two constants:

$$u(x) = a_1 + a_2 x \tag{11}$$

The constants a_1 and a_2 are obtained from the values of $u(x)$ at the two nodes; that is,

$$\begin{aligned} \text{at } x = 0, &\qquad u_1 = a_1 \\ \text{at } x = L, &\qquad u_2 = a_1 + a_2 L \end{aligned} \tag{12}$$

Equations (12) allow a_1 and a_2 to be determined in terms of u_1 and u_2 and substituted into (11) to give

$$u(x) = \left[1 - \frac{x}{L}, \frac{x}{L}\right]\begin{bmatrix} u_1 \\ u_2 \end{bmatrix} \tag{13}$$

This relation corresponds to (3). The polynomials in the N matrix, in this case $1 - x/L$ and x/L, are called the shape functions.

Since

$$\sigma = \sigma_x = E\varepsilon_x = E\varepsilon \tag{14}$$

with

$$\varepsilon = \frac{\partial u}{\partial x} \tag{15}$$

the matrix D reduces to the scalar E. Using (13), (15) becomes

$$\varepsilon = \left[-\frac{1}{L}, \frac{1}{L}\right]\begin{bmatrix} u_1 \\ u_2 \end{bmatrix} \tag{16}$$

which corresponds to (4) and identifies B as

$$B = \left[-\frac{1}{L}, \frac{1}{L}\right] \tag{17}$$

Substituting (17) into (7) gives

$$U - \frac{1}{2}\begin{bmatrix} u_1 \\ u_2 \end{bmatrix}^t \int_0^L ES \begin{bmatrix} \dfrac{1}{L^2} & -\dfrac{1}{L^2} \\[2mm] -\dfrac{1}{L^2} & \dfrac{1}{L^2} \end{bmatrix} dx \begin{bmatrix} u_1 \\ u_2 \end{bmatrix} \tag{18}$$

$$= \frac{1}{2}\frac{ES}{L}\begin{bmatrix} u_1 \\ u_2 \end{bmatrix}^t \begin{bmatrix} 1 & -1 \\ -1 & 1 \end{bmatrix}\begin{bmatrix} u_1 \\ u_2 \end{bmatrix} \tag{19}$$

and the stiffness matrix of the element is then

$$k = \frac{ES}{L}\begin{bmatrix} 1 & -1 \\ -1 & 1 \end{bmatrix} \tag{20}$$

This matrix is singular, i.e. its determinant is zero. This is due to the fact that the displacement function (11) allows rigid-body translation.

Beam in bending

An element of the beam is shown in Figure 2. It has two nodes with two degrees of freedom at each node; namely v, the lateral displacement, and ψ, the slope. In what follows, only the case of a Bernoulli–Euler beam is considered. This means that the secondary effects of transverse shear and rotatory inertia are neglected.

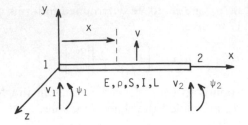

Figure 2   A beam in bending

The nodal displacement vector is

$$\delta = [v_1, \psi_1, v_2, \psi_2]^t \tag{21}$$

with

$$\psi = \frac{\partial v}{\partial x} \tag{22}$$

Since there are four nodal displacements, the displacement hypothesis is chosen to have four constants:

$$v = a_1 + a_2 x + a_3 x^2 + a_4 x^3 \tag{23}$$

Then from (22), the slope is

$$\psi = a_2 + 2a_3 x + 3a_4 x^2 \tag{24}$$

The constants a_1, a_2, a_3, a_4 are obtained from the values of v and ψ at the two nodes; that is,
   at $x = 0$:

$$v_1 = a_1$$
$$\psi_1 = a_2 \tag{25}$$

   at $x = L$:

$$v_2 = a_1 + a_2 L + a_3 L^2 + a_4 L^3$$
$$\psi_2 = a_2 + 2a_3 L + 3a_4 L^2 \tag{26}$$

The constants a_1, a_2, a_3, a_4 can be determined from (25) and (26) in terms of v_1, ψ_1, v_2, ψ_2. Substituting these constants into (23) gives

$$v(x) = \left[1 - \frac{3x^2}{L^2} + \frac{2x^3}{L^3}, \quad x - \frac{2x^2}{L} + \frac{x^3}{L^2}, \frac{3x^2}{L^2} - \frac{2x^3}{L^3}, \quad -\frac{x^2}{L} + \frac{x^3}{L^2}\right] \begin{bmatrix} v_1 \\ \psi_1 \\ v_2 \\ \psi_2 \end{bmatrix} \tag{27}$$

This equation corresponds to (3). For this case, the matrix D also reduces to a scalar equal to E. The bending stress and strain are related by (14), where now

$$\varepsilon = \varepsilon_x = -y \frac{\partial^2 v}{\partial x^2} \tag{28}$$

Using (27), ε becomes

$$\varepsilon = -y \left[-\frac{6}{L^2} + \frac{12x}{L^3}, \quad -\frac{4}{L} + \frac{6x}{L^2}, \frac{6}{L^2} - \frac{12x}{L^3}, \quad -\frac{2}{L} + \frac{6x}{L^2} \right] \begin{bmatrix} v_1 \\ \psi_1 \\ v_2 \\ \psi_2 \end{bmatrix} \tag{29}$$

This equation corresponds to (4) and identifies B. Using this expression for B in (7) and completing all the calculations gives

$$U = \frac{1}{2} \frac{EI}{L^3} \begin{bmatrix} v_1 \\ \psi_1 \\ v_2 \\ \psi_2 \end{bmatrix}^t \begin{bmatrix} 12 & 6L & -12 & 6L \\ 6L & 4L^2 & -6L & 2L^2 \\ -12 & -6L & 12 & -6L \\ 6L & 2L^2 & -6L & 4L^2 \end{bmatrix} \begin{bmatrix} v_1 \\ \psi_1 \\ v_2 \\ \psi_2 \end{bmatrix} \tag{30}$$

and the stiffness matrix of the element is then

$$k = \frac{EI}{L^3} \begin{bmatrix} 12 & 6L & -12 & 6L \\ 6L & 4L^2 & -6L & 2L^2 \\ -12 & -6L & 12 & -6L \\ 6L & 2L^2 & -6L & 4L^2 \end{bmatrix} \tag{31}$$

This matrix is singular and of rank two; this means that only two of the vectors which compose k are linearly independent. This is because the displacement function (23) allows two rigid-body motions, one in translation and one in rotation.

5.1.3   Kinetic energy – The mass matrix

The general expression for kinetic energy of an element is

$$T = \frac{1}{2} \int_\tau \rho V^2 \, d\tau \tag{32}$$

where $V = d^\circ$. Using (3), the velocity V becomes

$$V = N\delta^\circ \tag{33}$$

Substituting this into (32) gives

$$T = \frac{1}{2} \int_\tau \rho (N\delta^\circ)^t N\delta^\circ \, d\tau \tag{34}$$

which can be written in the form

$$T = \tfrac{1}{2}\delta^{\text{ot}} m \delta^{\circ} \tag{35}$$

where

$$m = \int_{\tau} \rho N^{\text{t}} N \, \text{d}\tau \tag{36}$$

is a symmetric mass matrix. The mass matrix defined by (36) is called the consistent mass matrix because it is calculated using the same shape functions as the stiffness matrix; in this sense, the calculations of k and m are consistent.

Instead of a consistent mass matrix, a lumped mass matrix m_1 is often used. This matrix is obtained by assuming that the mass of the element is concentrated, or lumped, at the nodes of the element. The advantage of lumping the mass is that the mass matrix becomes easy to construct and is diagonal. A diagonal mass matrix is convenient because it conserves computer memory and simplifies the algorithms used to compute frequencies and modes. For several elements, such as a bar in longitudinal motion, it is observed that the lumped mass matrix m_1 is as efficient as the consistent mass matrix m.

Bar in longitudinal motion

From (13):

$$N = \left[1 - \frac{x}{L}, \frac{x}{L} \right] \tag{37}$$

and combining this with (34) gives

$$T = \tfrac{1}{2}\rho S \begin{bmatrix} u_1^{\circ} \\ u_2^{\circ} \end{bmatrix}^{\text{t}} \int_0^L \begin{bmatrix} \left(1 - \dfrac{x}{L}\right)^2 & \dfrac{x}{L}\left(1 - \dfrac{x}{L}\right) \\ \dfrac{x}{L}\left(1 - \dfrac{x}{L}\right) & \left(\dfrac{x}{L}\right)^2 \end{bmatrix} \text{d}x \begin{bmatrix} u_1^{\circ} \\ u_2^{\circ} \end{bmatrix} \tag{38}$$

$$= \frac{1}{2}\frac{\rho SL}{6} \begin{bmatrix} u_1^{\circ} \\ u_2^{\circ} \end{bmatrix}^{\text{t}} \begin{bmatrix} 2 & 1 \\ 1 & 2 \end{bmatrix} \begin{bmatrix} u_1^{\circ} \\ u_2^{\circ} \end{bmatrix} \tag{39}$$

Hence the consistent mass matrix is

$$m = \frac{\rho SL}{6} \begin{bmatrix} 2 & 1 \\ 1 & 2 \end{bmatrix} \tag{40}$$

The lumped mass matrix m_1 is obtained as follows:

$$T = \frac{1}{2}\frac{\rho SL}{2}(u_1^{\circ 2} + u_2^{\circ 2}) \tag{41}$$

$$= \frac{1}{2}\frac{\rho SL}{2} \begin{bmatrix} u_1^{\circ 2} \\ u_2^{\circ 2} \end{bmatrix}^{\text{t}} \begin{bmatrix} 1 & 0 \\ 0 & 1 \end{bmatrix} \begin{bmatrix} u_1^{\circ} \\ u_2^{\circ} \end{bmatrix} \tag{42}$$

Hence

$$m_1 = \frac{\rho SL}{2} \begin{bmatrix} 1 & 0 \\ 0 & 1 \end{bmatrix} \tag{43}$$

Beam in bending

Using the shape functions given by (27) in expression (34) and completing all the calculations, results in:

$$T = \frac{1}{2} \frac{\rho SL}{420} \begin{bmatrix} v_1^\circ \\ \psi_1^\circ \\ v_2^\circ \\ \psi_2^\circ \end{bmatrix}^t \begin{bmatrix} 156 & 22L & 54 & -13L \\ 22L & 4L^2 & 13L & -3L^2 \\ 54 & 13L & 156 & -22L \\ -13L & -3L^2 & -22L & 4L^2 \end{bmatrix} \begin{bmatrix} v_1^\circ \\ \psi_1^\circ \\ v_2^\circ \\ \psi_2^\circ \end{bmatrix} \tag{44}$$

and, by inspection, the consistent mass matrix is

$$m = \frac{\rho SL}{420} \begin{bmatrix} 156 & 22L & 54 & -13L \\ 22L & 4L^2 & 13L & -3L^2 \\ 54 & 13L & 156 & -22L \\ -13L & -3L^2 & -22L & 4L^2 \end{bmatrix} \tag{45}$$

The lumped mass matrix is obtained from

$$T = \frac{1}{2} \frac{\rho SL}{2} (v_1^{\circ 2} + v_2^{\circ 2}) \tag{46}$$

Hence

$$m_1 = \frac{\rho SL}{2} \begin{bmatrix} 1 & 0 & 0 & 0 \\ 0 & 0 & 0 & 0 \\ 0 & 0 & 1 & 0 \\ 0 & 0 & 0 & 0 \end{bmatrix} \tag{47}$$

In bending, the consistent mass matrix is more efficient than the lumped mass matrix because the latter does not take into account the effect of beam element rotation.

5.1.4 Assembly

To assemble element properties into structure properties it is necessary to write that the kinetic energy and the strain energy of the structure are the sum of the kinetic energy and strain energy of each of the elements. The generalized forces are obtained from virtual work. Lagrange's equations are then used to give the classical form of the differential equations of motion for the whole structure; that is,

$$M\delta^{\circ\circ} + K\delta = F(t) \tag{48}$$

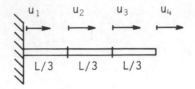

Figure 3 Clamped–free bar modeled by three finite elements

where the structure mass matrix M is symmetric, positive definite, and the structure stiffness matrix K is symmetric, positive semi-definite. The vectors δ and $F(t)$ contain all the nodal displacements and generalized forces. In general, the mass and stiffness matrices have a banded form.

The process of assembly is illustrated by considering a fixed–free bar of length L undergoing longitudinal motion and which has been discretized into three finite elements of the type shown in Figure 1. As shown in Figure 3, the bar has a force $F(t)$ applied to node 3. Summing the elements strain energies (19), and accounting for the boundary condition $u_1 = 0$, gives

$$U = \frac{1}{2}\frac{3ES}{L}\begin{bmatrix}0\\u_2\end{bmatrix}^t\begin{bmatrix}1 & -1\\-1 & 1\end{bmatrix}\begin{bmatrix}0\\u_2\end{bmatrix} + \frac{1}{2}\frac{3ES}{L}\begin{bmatrix}u_2\\u_3\end{bmatrix}^t\begin{bmatrix}1 & -1\\-1 & 1\end{bmatrix}\begin{bmatrix}u_2\\u_3\end{bmatrix}$$
$$+\frac{1}{2}\frac{3ES}{L}\begin{bmatrix}u_3\\u_4\end{bmatrix}^t\begin{bmatrix}1 & -1\\-1 & 1\end{bmatrix}\begin{bmatrix}u_3\\u_4\end{bmatrix} \tag{49}$$

Applying Lagrange's equations gives

$$\begin{bmatrix}\dfrac{\partial U}{\partial u_2}\\[2mm]\dfrac{\partial U}{\partial u_3}\\[2mm]\dfrac{\partial U}{\partial u_4}\end{bmatrix} = \frac{3ES}{L}\begin{bmatrix}2 & -1 & 0\\-1 & 2 & -1\\0 & -1 & 1\end{bmatrix}\begin{bmatrix}u_2\\u_3\\u_4\end{bmatrix} = K\delta \tag{50}$$

It can be observed that the structure stiffness matrix K in (50) can be obtained by a superposition of the three element stiffness matrices k given by (20) provided L is replaced by $L/3$, and the rows and columns corresponding to $u_1 = 0$ are deleted.

Similarly, using the consistent mass matrix formulation of (39) gives

$$T = \frac{1}{2}\frac{\rho SL}{18}\begin{bmatrix}0\\u_2{}^\circ\end{bmatrix}^t\begin{bmatrix}2 & 1\\1 & 2\end{bmatrix}\begin{bmatrix}0\\u_2{}^\circ\end{bmatrix} + \frac{1}{2}\frac{\rho SL}{18}\begin{bmatrix}u_2{}^\circ\\u_3{}^\circ\end{bmatrix}^t\begin{bmatrix}2 & 1\\1 & 2\end{bmatrix}\begin{bmatrix}u_2{}^\circ\\u_3{}^\circ\end{bmatrix}$$
$$+\frac{1}{2}\frac{\rho SL}{18}\begin{bmatrix}u_3{}^\circ\\u_4{}^\circ\end{bmatrix}^t\begin{bmatrix}2 & 1\\1 & 2\end{bmatrix}\begin{bmatrix}u_3{}^\circ\\u_4{}^\circ\end{bmatrix} \tag{51}$$

and application of Lagrange's equations gives

$$
\begin{bmatrix} \dfrac{d}{dt}\left(\dfrac{\partial T}{\partial u_2{}^\circ}\right) \\[2mm] \dfrac{d}{dt}\left(\dfrac{\partial T}{\partial u_3{}^\circ}\right) \\[2mm] \dfrac{d}{dt}\left(\dfrac{\partial T}{\partial u_4{}^\circ}\right) \end{bmatrix} = \frac{\rho SL}{18} \begin{bmatrix} 4 & 1 & 0 \\ 1 & 4 & 1 \\ 0 & 1 & 2 \end{bmatrix} \begin{bmatrix} u_2{}^{\circ\circ} \\ u_3{}^{\circ\circ} \\ u_4{}^{\circ\circ} \end{bmatrix} = M\delta^{\circ\circ} \tag{52}
$$

where, again, M can be obtained by a superposition of the element mass matrices m.

Using (42) and recalculating (51) for the lumped mass matrix formulation gives

$$
T = \frac{\rho SL}{6} u_2{}^{\circ 2} + \frac{\rho SL}{6} u_3{}^{\circ 2} + \frac{\rho SL}{12} u_4{}^{\circ 2} \tag{53}
$$

from which the structure lumped mass matrix M_1 can be shown to be

$$
M_1 = \frac{\rho SL}{6} \begin{bmatrix} 2 & 0 & 0 \\ 0 & 2 & 0 \\ 0 & 0 & 1 \end{bmatrix} \tag{54}
$$

The force $F(t)$ acting at mode 3 results in the external force vector

$$
[0, F(t), 0]^t \tag{55}
$$

In summary, this section has presented systematic ways of deriving and assembling the stiffness and mass matrices of simple elements. The library of finite elements available in many structural dynamics computer programs is much more extensive than can be presented here. It is also worth pointing out the connection between the Rayleigh–Ritz method which was used in Chapters 3 and 4 and the finite element method used in this chapter. The Rayleigh–Ritz method uses a single displacement hypothesis which spans the entire structure; the finite element method uses many similar displacements functions, each being defined over a different element. In a broad sense, the finite element method is a piecewise version of the Rayleigh–Ritz method.

5.2 Frequencies and Modes; Response to an Excitation

The frequencies and modes are obtained from the homogeneous equation of motion:

$$
M\delta^{\circ\circ} + K\delta = 0 \tag{56}
$$

Let

$$
\delta = \delta_0 e^{j\omega t} \tag{57}
$$

then (56) becomes

$$
\omega^2 M\delta_0 = K\delta_0 \tag{58}
$$

This is the classical algebraic eigenvalue–eigenvector problem discussed in Chapter 3.

For a finite element formulation of (56), the matrices M and K can have several hundreds to several thousands degrees of freedom. Extracting the frequencies and modes from such a large system requires computer methods which are specifically adapted to minimize the amount of execution time and the size of the required core memory. The details of such specific eigenvalue algorithms are too specialized for this text. However, a discussion will be given of two aspects which are particularly useful in the numerical calculation of large systems.

Rigid-body modes

In order to avoid problems due to zero values of frequency, it is useful to associate with equation (58) the identity

$$aM\delta_0 = aM\delta_0 \tag{59}$$

Combining (58) and (59) gives

$$(\omega^2 + a)M\delta_0 = (K + aM)\delta_0 \tag{60}$$

which can be written

$$\omega^{*2}M\delta_0 = K^*\delta_0 \tag{61}$$

The frequencies found from (61) are not zero, since

$$
\begin{aligned}
\omega_1^{*2} &= \omega_1^2 + a = a \\
&\vdots \\
\omega_i^{*2} &= \omega_i^2 + a \\
&\vdots
\end{aligned}
\tag{62}
$$

The modes associated with ω_i^* are the same as those associated with ω_i of the initial system. The constant a is positive and must be choosen with equation (62) in mind: that is, $\omega_i^2 + a$ must not be $\simeq \omega_i^2$ or $\simeq a$.

Diagonal mass matrix

Let k_{ij} be the elements of K, and m_{ii} the elements of M where M is diagonal (i.e. $m_{ij} = 0$, $i \neq j$). The standard form of (58) is

$$\omega^2 \delta_0 = A\delta_0 \tag{63}$$

where $A = M^{-1}K$. The matrix A is not symmetric and this complicates the determination of the eigenvalues and eigenvectors. In addition, A cannot be represented by its semi-bandwidth and this increases the size of core memory which must be used. These difficulties can be avoided by adopting the change of variable

$$\delta_i^* = \sqrt{m_{ii}}\, \delta_i \tag{64}$$

which allows (56) to be rewritten as

$$I\delta^{*\infty} + K^*\delta^* = 0 \qquad (65)$$

where I is the unit matrix and

$$k_{ij}^* = \frac{k_{ij}}{\sqrt{m_{ii}m_{jj}}} \qquad (66)$$

Clearly, $A^* = I^{-1}K^* = K^*$ and is symmetric.

Response to an excitation

In determining the response for a system which has been discretized by finite elements, one proceeds as in Chapter 3; that is, one uses a modal or pseudo-modal method or, in the case of an arbitrary excitation, a step-by-step method which can, if necessary, be combined with the modal or pseudo-modal method.

5.3 Structural Modifications

In practice it is useful to have a method of calculating how minor changes in the components of a structure affect the frequencies of the whole structure without being obliged to recalculate the new frequencies. The finite element method can provide such a method and thereby allows economies in manpower and computer time.

5.3.1 Influence of a small structural modification

Let ω_i, ϕ_i, $\phi_i^t M \phi_i$, $\phi_i^t K \phi_i$ be the ith frequency, mode, modal mass, and modal stiffness of the unmodified structure. Then one has

$$\omega_i^2 \phi_i^t M \phi_i = \phi_i^t K \phi_i \qquad (67)$$

After modification, the mass matrix becomes $M + \Delta M$ and the stiffness matrix becomes $K + \Delta K$. The frequencies and modes become $\omega_i + \Delta\omega_i$ and $\phi_i + \Delta\phi_i$ and, instead of (67), one has

$$(\omega_i + \Delta\omega_i)^2(\phi_i + \Delta\phi_i)^t(M + \Delta M)(\phi_i + \Delta\phi_i) = (\phi_i + \Delta\phi_i)^t(K + \Delta K)(\phi_i + \Delta\phi_i) \qquad (68)$$

Expanding (68) and keeping only first-order terms gives

$$\Delta\omega_i = \frac{\omega_i}{2}\frac{\phi_i^t \Delta K \phi_i}{\phi_i^t K \phi_i} - \frac{\omega_i}{2}\frac{\phi_i^t \Delta M \phi_i}{\phi_i^t M \phi_i} \qquad (69)$$

The same expression may be found by using Rayleigh's method rather than starting with (67). To see this, suppose that the modes are not significantly changed when M and K are modified. Then for mode ϕ_i:

$$T_i = \tfrac{1}{2}\phi_i^t(M+\Delta M)\phi_i p_i^{\circ 2} \tag{70}$$

$$U_i = \tfrac{1}{2}\phi_i^t(K+\Delta K)\phi_i p_i^2 \tag{71}$$

and Rayleigh's method leads to

$$\omega_i^{*2} = \frac{\phi_i^t K\phi_i + \phi_i^t \Delta K\phi_i}{\phi_i^t M\phi_i + \phi_i^t \Delta M\phi_i} \tag{72}$$

which can be written

$$\omega_i^{*2} = \omega_i^2 \frac{1+\phi_i^t \Delta K\phi_i/\phi_i^t K\phi_i}{1+\phi_i^t \Delta M\phi_i/\phi_i^t M\phi_i} \tag{73}$$

or

$$\omega_i^{*} \simeq \omega_i + \frac{\omega_i}{2}\frac{\phi_i^t \Delta K\phi_i}{\phi_i^t K\phi_i} - \frac{\omega_i}{2}\frac{\phi_i^t \Delta M\phi_i}{\phi_i^t M\phi_i} \tag{74}$$

and, as $\omega_i^{*} - \omega_i = \Delta\omega_i$, equation (69) is exactly recovered.

In using (69) or (74), it is only necessary to know ΔK and ΔM and to compute the matrix products $\phi_i^t \Delta K\phi_i$ and $\phi_i^t \Delta M\phi_i$. These operations can be performed for many different structural modifications without the expense of calculating the frequencies of the modified structure.

5.3.2 How to modify a structure

It is very difficult to anticipate the influence of a particular modification on the dynamic behavior of a complex structure. In the process of analyzing a structure for its frequencies and modes, it is easy to determine the quantities $\phi_i^t M\phi_i$ and $\phi_i^t K\phi_i$ for each finite element. The kinetic and strain energy corresponding to a mode ω_i, ϕ_i can then be written as a sum over the N elements, i.e.

$$\begin{aligned} T_i &= \tfrac{1}{2}\phi_i^t M\phi_i p_i^{\circ 2} \\ &= \frac{1}{2}\sum_{r=1}^{N}(\phi_{ir}^t m_r\phi_{ir})p_i^{\circ 2} \end{aligned} \tag{75}$$

$$\begin{aligned} U_i &= \tfrac{1}{2}\phi_i^t K\phi_i p_i^2 \\ &= \frac{1}{2}\sum_{r=1}^{N}(\phi_{ir}^t k_r\phi_{ir})p_i^2 \end{aligned} \tag{76}$$

where k_r and m_r are the stiffness and mass matrices of element r and ϕ_{ir} is the portion of the vector ϕ_i associated with the motion of the rth element.

It is sufficient to examine the relative importance of these terms for a mode to establish the influence of a particular finite element or group of finite elements on the behavior of the whole structure. It is then possible to choose a good location for a structural modification such as local mass or stiffness, or even a layer of damping material. Another important part of finite element analysis of structures which have a large number of degrees of freedom is the judgment as to whether the answers for frequencies and modes are correct. With the vast quantity of computer output in such cases, it is very difficult to use physical intuition to validate the answers. Ways have to be found to condense the output data so that its validity can be established by intuition or by simplified analysis. One of the most effective ways of doing this is to calculate and display the distribution of element kinetic and strain energy. Such displays provide concise information about motion and stress distribution and facilitate judgments about correctness.

5.4 The Method of Substructures

This method is used to predict the dynamic behavior of a complex structure by dividing the structure into a series of smaller structures, called substructures, and studying the dynamic characteristics of these components. Substructuring allows a considerable reduction in the number of degrees of freedom which would otherwise be required to model the entire structure. This method can also be extended so that some substructures are studied analytically and some experimentally; however, this refinement will not be discussed here. In order to simplify the following presentation, it is supposed that the structure is divided into just two substructures; see Figure 4. The first substructure SS1 is connected to the second substructure SS2 by common degrees of freedom denoted by δ_c. Each substructure has, in addition, its own internal degrees of freedom denoted by $\delta_i^{(1)}$ and $\delta_i^{(2)}$. The same approach as described in section 5.1 is used to obtain the equations of motion. It is first necessary to write the kinetic energy and strain energy of the substructures and to write the virtual work of their external forces. Let $F^{(1)}$, $F^{(2)}$ be the external forces acting on the internal degrees of freedom $\delta_i^{(1)}$, $\delta_i^{(2)}$ and assume, for simplification, that there are no external forces

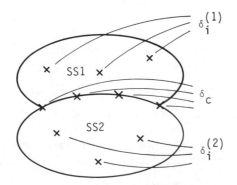

Figure 4   Structure decomposed into
two substructures

associated with δ_c. Using the additive property of the energies:

$$T = T^{(1)} + T^{(2)}$$
$$U = U^{(1)} + U^{(2)}$$

(77)

Adopting the finite element formulation of T and U and partitioning the matrices to separate the internal degrees of freedom $\delta_i^{(1)}$, $\delta_i^{(2)}$ from the connection degrees of freedom δ_c gives

$$T^{(1)} = \frac{1}{2} \begin{bmatrix} \delta_i^{\circ(1)} \\ \delta_c^{\circ} \end{bmatrix}^t \begin{bmatrix} m_{ii}^{(1)} & m_{ic}^{(1)} \\ m_{ci}^{(1)} & m_{cc}^{(1)} \end{bmatrix} \begin{bmatrix} \delta_i^{\circ(1)} \\ \delta_c^{\circ} \end{bmatrix}$$

(78)

$$U^{(1)} = \frac{1}{2} \begin{bmatrix} \delta_i^{(1)} \\ \delta_c \end{bmatrix}^t \begin{bmatrix} k_{ii}^{(1)} & k_{ic}^{(1)} \\ k_{ci}^{(1)} & k_{cc}^{(1)} \end{bmatrix} \begin{bmatrix} \delta_i^{(1)} \\ \delta_c \end{bmatrix}$$

(79)

$$T^{(2)} = \frac{1}{2} \begin{bmatrix} \delta_i^{\circ(2)} \\ \delta_c^{\circ} \end{bmatrix}^t \begin{bmatrix} m_{ii}^{(2)} & m_{ic}^{(2)} \\ m_{ci}^{(2)} & m_{cc}^{(2)} \end{bmatrix} \begin{bmatrix} \delta_i^{\circ(2)} \\ \delta_c^{\circ} \end{bmatrix}$$

(80)

$$U^{(2)} = \frac{1}{2} \begin{bmatrix} \delta_i^{(2)} \\ \delta_c \end{bmatrix}^t \begin{bmatrix} k_{ii}^{(2)} & k_{ic}^{(2)} \\ k_{ci}^{(2)} & k_{cc}^{(2)} \end{bmatrix} \begin{bmatrix} \delta_i^{(2)} \\ \delta_c \end{bmatrix}$$

(81)

Different variations of the method of substructures are characterized by the changes in variable introduced into (78), (79), (80), (81), and the virtual work expression. Two frequently used methods are presented below.

5.4.1 Method using free modes

In this method, the vibration modes ϕ of each substructure are calculated supposing the connection nodes are free. The kinetic and strain energies and the virtual work of the external forces are expressed in terms of these free modes. The corresponding variables are:

$q^{(1)}$ SS1 generalized coordinates
$q^{(2)}$ SS2 generalized coordinates
δ_c the physical coordinates connecting SS1 to SS2.

Using (78) and (79), substructure 1, which is free at its boundaries and without external forces, has a dynamic behavior defined by

$$\begin{bmatrix} m_{ii}^{(1)} & m_{ic}^{(1)} \\ m_{ci}^{(1)} & m_{cc}^{(1)} \end{bmatrix} \begin{bmatrix} \delta_i^{\circ\circ(1)} \\ \delta_c^{\circ\circ} \end{bmatrix} + \begin{bmatrix} k_{ii}^{(1)} & k_{ic}^{(1)} \\ k_{ci}^{(1)} & k_{cc}^{(1)} \end{bmatrix} \begin{bmatrix} \delta_i^{(1)} \\ \delta_c \end{bmatrix} = 0$$

(82)

The solution of this system supplies a set of modes which can be assembled into the modal matrix $\phi^{(1)}$ and this then gives

$$\begin{bmatrix} \delta_i^{(1)} \\ \delta_c \end{bmatrix} = \begin{bmatrix} \phi_i^{(1)} \\ \phi_c^{(1)} \end{bmatrix} q^{(1)}$$

(83)

According to (83),

$$\delta_i^{(1)} = \phi_i^{(1)} q^{(1)} \tag{84}$$

and substituting this into (78) and (79) gives

$$T^{(1)} = \frac{1}{2} \begin{bmatrix} q^{o(1)} \\ \delta_c^o \end{bmatrix}^t \begin{bmatrix} \phi_i^{(1)t} m_{ii}^{(1)} \phi_i^{(1)} & \phi_i^{(1)t} m_{ic}^{(1)} \\ m_{ci}^{(1)} \phi_i^{(1)} & m_{cc}^{(1)} \end{bmatrix} \begin{bmatrix} q^{o(1)} \\ \delta_c^o \end{bmatrix} \tag{85}$$

$$U^{(1)} = \frac{1}{2} \begin{bmatrix} q^{(1)} \\ \delta_c \end{bmatrix}^t \begin{bmatrix} \phi_i^{(1)t} k_{ii}^{(1)} \phi_i^{(1)} & \phi_i^{(1)t} k_{ic}^{(1)} \\ k_{ci}^{(1)} \phi_i^{(1)} & k_{cc}^{(1)} \end{bmatrix} \begin{bmatrix} q^{(1)} \\ \delta_c \end{bmatrix} \tag{86}$$

The energies $T^{(2)}$ and $U^{(2)}$ will have expressions identical to (85) and (86), except superscript (1) is replaced by (2).

Since here $F_c = 0$, the force vector corresponding to $\delta_i^{(1)}$, $\delta_i^{(2)}$, δ_c is

$$[F^{(1)}, F^{(2)}, 0]^t \tag{87}$$

In the light of (84) and of a similar expression for SS2, the generalized force vector corresponding to $q^{(1)}$, $q^{(2)}$, δ_c is

$$[\phi_i^{(1)t} F^{(1)}, \phi_i^{(2)t} F^{(2)}, 0]^t \tag{88}$$

Applying Lagrange's equations to the structure then gives

$$\begin{bmatrix} \phi_i^{(1)t} m_{ii}^{(1)} \phi_i^{(1)} & 0 & \phi_i^{(1)t} m_{ic}^{(1)} \\ 0 & \phi_i^{(2)t} m_{ii}^{(2)} \phi_i^{(2)} & \phi_i^{(2)t} m_{ic}^{(2)} \\ m_{ci}^{(1)} \phi_i^{(1)} & m_{ci}^{(2)} \phi_i^{(2)} & m_{cc}^{(1)} + m_{cc}^{(2)} \end{bmatrix} \begin{bmatrix} q^{oo(1)} \\ q^{oo(2)} \\ \delta_c^{oo} \end{bmatrix} +$$

$$\begin{bmatrix} \phi_i^{(1)t} k_{ii}^{(1)} \phi_i^{(1)} & 0 & \phi_i^{(1)t} k_{ic}^{(1)} \\ 0 & \phi_i^{(2)t} k_{ii}^{(2)} \phi_i^{(2)} & \phi_i^{(2)t} k_{ic}^{(2)} \\ k_{ci}^{(1)} \phi_i^{(1)} & k_{ci}^{(2)} \phi_i^{(2)} & k_{cc}^{(1)} + k_{cc}^{(2)} \end{bmatrix} \begin{bmatrix} q^{(1)} \\ q^{(2)} \\ \delta_c \end{bmatrix} = \begin{bmatrix} \phi_i^{(1)t} F^{(1)} \\ \phi_i^{(2)t} F^{(2)} \\ 0 \end{bmatrix} \tag{89}$$

The system (89) is now the analytical model of the entire structure. Notice that all the necessary quantities to form (89) are obtained from dynamic analysis of the two substructures. In other words, the behavior of the entire structure has been obtained in terms of the behavior of each substructure. The solution of (89) is obtained by the usual methods for determining system frequencies or system response. The recovery of the physical variable is accomplished via (84) and its analog for SS2. The number of degrees of freedom can be reduced only in one substructure; see exercise 14.

5.4.2 Method using constrained modes

This method is only presented briefly in order to avoid a too lengthy discussion of the substructure method. The connection degrees of freedom δ_c are retained as variables and the internal displacement variables are determined following two steps.

1. Static displacement: Let $F^{(1)} = 0$, then for SS1 under static conditions:

$$\begin{bmatrix} k_{ii}^{(1)} & k_{ic}^{(1)} \\ k_{ci}^{(1)} & k_{cc}^{(1)} \end{bmatrix} \begin{bmatrix} \delta_i^{(1)} \\ \delta_c \end{bmatrix} = 0 \tag{90}$$

from which one can obtain the relation

$$\delta_i^{(1)} = -(k_{ii}^{(1)})^{-1} \cdot k_{ic}^{(1)} \cdot \delta_c \tag{91}$$

2. Dynamic displacements: In this method the connection nodes are constrained by clamping (i.e. $\delta_c = 0$); also it is assumed there are no external forces $(F^{(1)} = F^{(2)} = 0)$. The dynamic behavior of SS1 is then found from

$$m_{ii}^{(1)} \delta_i^{(1)\infty} + k_{ii}^{(1)} \delta_i^{(1)} = 0 \tag{92}$$

The modes of (92) give

$$\delta_i^{(1)} = \phi_i^{(1)} q^{(1)} \tag{93}$$

Combining equations (91) and (93) gives the matrix relation

$$\begin{bmatrix} \delta_i^{(1)} \\ \delta_c \end{bmatrix} = \begin{bmatrix} \phi_i^{(1)} & -(k_{ii}^{(1)})^{-1} k_{ic}^{(1)} \\ 0 & I \end{bmatrix} \begin{bmatrix} q^{(1)} \\ \delta_c \end{bmatrix} \tag{94}$$

where I is the identity matrix.

It is now possible to substitute (94) into (78) and (79) to obtain $T^{(1)}$ and $U^{(1)}$ and one can proceed in a similar fashion to obtain $T^{(2)}$ and $U^{(2)}$.

The generalized forces vector becomes

$$\begin{bmatrix} \phi_i^{(1)t} F^{(1)} \\ \phi_i^{(2)t} F^{(2)} \\ -[(k_{ii}^{(1)})^{-1} k_{ic}^{(1)}]^t F^{(1)} - [(k_{ii}^{(2)})^{-1} k_{ic}^{(2)}]^t F^{(2)} \end{bmatrix} \tag{95}$$

All the necessary terms to formulate the equations of motion in terms of $q^{(1)}$, $q^{(2)}$, δ_c are now available. After solution, the physical coordinates for SS1 can be recovered from (94) and those for SS2 from (94) with superscript (2) exchanged for (1).

5.5 Exercises

Exercise 1: *A beam is simply supported at each end; see Figure 5. Using only a single finite element, calculate its lowest frequency and the corresponding mode.*

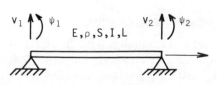

Figure 5

The consistent mass matrix has to be used, otherwise the kinetic energy would be zero. The equation of motion is

$$\frac{\rho SL}{420}\begin{bmatrix} 4L^2 & -3L^2 \\ -3L^2 & 4L^2 \end{bmatrix}\begin{bmatrix} \ddot{\psi}_1 \\ \ddot{\psi}_2 \end{bmatrix} + \frac{EI}{L^3}\begin{bmatrix} 4L^2 & 2L^2 \\ 2L^2 & 4L^2 \end{bmatrix}\begin{bmatrix} \psi_1 \\ \psi_2 \end{bmatrix} = 0$$

from which one finds

$$\omega = \frac{10.95}{L^2}\sqrt{\frac{EI}{\rho S}} \quad \text{and} \quad \begin{bmatrix} \psi_1 \\ \psi_2 \end{bmatrix} = \begin{bmatrix} 1 \\ -1 \end{bmatrix}$$

Using equation (27):

$$v(x) = \left(x - \frac{2x^2}{L} + \frac{x^3}{L^2}\right)(1) + \left(-\frac{x^2}{L} + \frac{x^3}{L^2}\right)(-1) = x - \frac{x^2}{L}$$

The above value of ω is within 11% of the exact fundamental frequency.

Exercise 2: *Repeat exercise 1 for a clamped–free cantilever beam. Use a consistent mass matrix.*

The frequency is

$$\omega = \frac{3.533}{L^2}\sqrt{\frac{EI}{\rho S}}$$

and the corresponding mode is

$$\begin{bmatrix} v_2 \\ \psi_2 \end{bmatrix} = \begin{bmatrix} 1 \\ \dfrac{1.378}{L} \end{bmatrix}$$

from which

$$v(x) = 1.622\frac{x^2}{L^2} - 0.622\frac{x^3}{L^3}$$

The above frequency is within 1% of the exact value.

Exercise 3: *Consider the same system as in exercise 1 of Chapter 4. Model the beam by three finite elements; see Figure 6. Calculate T and U using the consistent mass matrix. Also calculate the two lowest frequencies and modes.*

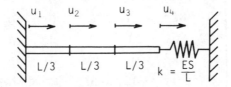

Figure 6

T is the same as given in this chapter in section 5.1.4. To obtain U it is necessary to add the strain energy of the spring; namely

$$U_s = \tfrac{1}{2}ku_4^2$$

Completing all the calculations results in the stiffness matrix

$$K = \frac{3ES}{L}\begin{bmatrix} 2 & -1 & 0 \\ -1 & 2 & -1 \\ 0 & -1 & \tfrac{4}{3} \end{bmatrix}$$

The frequencies and modes are

$$\omega_1 = \frac{2.068}{L}\sqrt{\frac{E}{\rho}}; \qquad \phi_1 = \begin{bmatrix} 1 \\ 1.560 \\ 1.433 \end{bmatrix}$$

$$\omega_2 = \frac{5.474}{L}\sqrt{\frac{E}{\rho}}; \qquad \phi_2 = \begin{bmatrix} 1 \\ -0.1410 \\ -0.9801 \end{bmatrix}$$

Exercise 4: *A cantilever beam has a length of 3L. Model the beam with three finite elements (see Figure 7) and write the system stiffness matrix and consistent mass matrix. Using program no. 5 of Chapter 7, calculate the first two frequencies, the corresponding modes, and the associated modal stiffnesses and masses.*

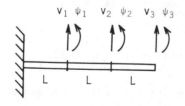

Figure 7

The solutions are:

$$K = \frac{EI}{L^3}\begin{bmatrix} 24 & 0 & -12 & 6L & 0 & 0 \\ 0 & 8L^2 & -6L & 2L^2 & 0 & 0 \\ -12 & -6L & 24 & 0 & -12 & 6L \\ 6L & 2L^2 & 0 & 8L^2 & -6L & 2L^2 \\ 0 & 0 & -12 & -6L & 12 & -6L \\ 0 & 0 & 6L & 2L^2 & -6L & 4L^2 \end{bmatrix}$$

$$M = \frac{\rho SL}{420} \begin{bmatrix} 312 & 0 & 54 & -13L & 0 & 0 \\ 0 & 8L^2 & 13L & -3L^2 & 0 & 0 \\ 54 & 13L & 312 & 0 & 54 & -13L \\ -13L & -3L^2 & 0 & 8L^2 & 13L & -3L^2 \\ 0 & 0 & 54 & 13L & 156 & -22L \\ 0 & 0 & -13L & -3L^2 & -22L & 4L^2 \end{bmatrix}$$

from which

$$\omega_1 = \frac{3.516}{(3L)^2} \sqrt{\frac{EI}{\rho S}}$$

$$\phi_1 = \left[1, \frac{1.821}{L}, 3.304, \frac{2.636}{L}, 6.041, \frac{2.772}{L} \right]^t$$

with

$$\phi_1^t M \phi_1 = 27.36 \rho SL$$

$$\phi_1^t K \phi_1 = 4.176 \frac{EI}{L^3}$$

and

$$\omega_2 = \frac{22.11}{(3L)^2} \sqrt{\frac{EI}{\rho S}}$$

$$\phi_2 = \left[1, \frac{0.9966}{L}, 0.7179, -\frac{1.672}{L}, -1.695, -\frac{2.704}{L} \right]^t$$

with

$$\phi_2^t M \phi_2 = 2.129 \rho SL$$

$$\phi_2^t K \phi_2 = 12.85 \frac{EI}{L^3}$$

The above frequencies are slightly above the exact values and their accuracy is very good.

Exercise 5: *Consider the same system as in exercise 4. In the present case, form the lumped mass matrix and use it to calculate the first two frequencies and modes. Use program no. 5,*

The solutions are:

$$M = \frac{\rho SL}{2} \begin{bmatrix} 2 & & & & & \\ & 0 & & & & \text{zero} \\ & & 2 & & & \\ & & & 0 & & \\ & & & & 1 & \\ \text{zero} & & & & & 0 \end{bmatrix}$$

from which

$$\omega_1 = \frac{3.346}{(3L)^2} \sqrt{\frac{EI}{\rho S}}$$

$$\phi_1 = \left[1, \frac{1.829}{L}, 3.339, \frac{2.700}{L}, 6.181, \frac{2.914}{L}\right]^t$$

and

$$\omega_2 = \frac{18.89}{(3L)^2} \sqrt{\frac{EI}{\rho S}}$$

$$\phi_2 = \left[1, \frac{1.058}{L}, 0.9673, -\frac{1.331}{L}, -1.368, -\frac{2.838}{L}\right]^t$$

These frequencies are lower than the exact value and their accuracy is not very good.

Exercise 6: *Consider the same system as in exercise* 23 *of Chapter* 4. *Model the beam by a single finite element (see Figure* 8) *and calculate the three frequencies and modes. Use the consistent mass matrix.*

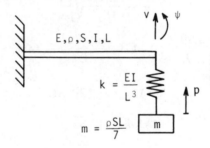

Figure 8

In this case, the kinetic and strain energies become

$$T = \frac{1}{2} \frac{\rho SL}{840} \begin{bmatrix} v^\circ \\ \psi^\circ \end{bmatrix}^t \begin{bmatrix} 156 & -22L \\ -22L & 4L^2 \end{bmatrix} \begin{bmatrix} v^\circ \\ \psi^\circ \end{bmatrix} + \frac{1}{2} \frac{\rho SL}{7} p^{\circ 2}$$

$$U = \frac{1}{2} \frac{EI}{L^3} \begin{bmatrix} v \\ \psi \end{bmatrix}^t \begin{bmatrix} 12 & -6L \\ -6L & 4L^2 \end{bmatrix} \begin{bmatrix} v \\ \psi \end{bmatrix} + \frac{1}{2} \frac{EI}{L^3} (v-p)^2$$

and application of Lagrange's equations gives

$$\rho SL \begin{bmatrix} 156/420 & -22L/420 & 0 \\ -22L/420 & 4L^2/420 & 0 \\ 0 & 0 & 1/7 \end{bmatrix} \begin{bmatrix} v^{\circ\circ} \\ \psi^{\circ\circ} \\ p^{\circ\circ} \end{bmatrix} + \frac{EI}{L^3} \begin{bmatrix} 13 & -6L & -1 \\ -6L & 4L^2 & 0 \\ -1 & 0 & 1 \end{bmatrix} \begin{bmatrix} v \\ \psi \\ p \end{bmatrix} = 0$$

from which the following results are obtained:

$$\omega_1 = \frac{2.143}{L^2} \sqrt{\frac{EI}{\rho S}}; \qquad \phi_1 = \begin{bmatrix} 0.3437 \\ 0.5004/L \\ 1 \end{bmatrix}$$

$$\omega_2 = \frac{4.346}{L^2} \sqrt{\frac{EI}{\rho S}}; \qquad \phi_2 = \begin{bmatrix} 0.7624 \\ 1/L \\ -0.4488 \end{bmatrix}$$

$$\omega_3 = \frac{34.92}{L^2} \sqrt{\frac{EI}{\rho S}}; \qquad \phi_3 = \begin{bmatrix} 0.1316 \\ 1/L \\ -0.0007 \end{bmatrix}$$

Exercise 7: *Use the system as in the preceding exercise. Prepare the data required by program no. 11 of Chapter 7. Calculate the first three frequencies and modes and determine the percentage distribution of the kinetic and strain energies for each element in each of the three modes. Use SI units with* $L = 2\ m$; $S = 10^{-2}\ m^2$; $I = 10^{-6}\ m^4$; $\rho = 7800\ kg/m^3$; $E = 2 \times 10^{11}\ N/m^2$.

Number of elements: 2
Number of nodes:3
Number of materials: 1
Number of sections: 1
Number of constrained nodes: 3

Kind of element 1: B
Kind of element 2: MSM

Beam element no. 1 – Number of 1st node: 1
Number of 2nd node: 2
Number of material: 1
Number of section: 1

Material no. 1 – Young's modulus: 2.E+11
Mass per unit volume: 7800

Section no. 1 – Area: 1.E−2
Inertia: 1.E−6

MSM Element no. 1 – Number of 1st node: 2
Mass in X: 0
Mass in Y: 0
Inertia in PSI: 0

System MSM no. 1 – Number of 2nd node: 3

                           Mass in X: 22.285714

                           Mass in Y: 0

                           Inertia in PSI: 0

                           Stiffness in X: 2.5E+4

                           Stiffness in Y: 0

                           Stiffness in PSI: 0

Node 1 – Absissa: 0

        Ordinate: 0

Node 2 – Absissa: 2

        Ordinate: 0

Node 3 – Absissa: 2

        Ordinate: −1 (This figure is only to establish a reference frame for the spring–mass system).

Number of 1st constrained node: 1

For this node condition in $u = 1$

For this node condition in $v = 1$

For this node condition in $PSI = 1$

Number of 2nd constrained node: 2

For this node condition in $u = 1$

For this node condition in $v = 0$

For this node condition in $PSI = 0$

Number of 3rd constrained node: 3

For this node condition in $u = 1$

For this node condition in $v = 0$

For this node condition in $PSI = 1$

    Results:

$$\omega_1 = 27.13 \text{ rad/sec}; \quad \phi_1 = [0.3437, 0.2502, 1]^t$$

$$\omega_3 = 55.02 \text{ rad/sec}; \quad \phi_2 = [1, 0.6558, -0.5886]^t$$

$$\omega_3 = 442.1 \text{ rad/sec}; \quad \phi_3 = [0.2631, 1, -0.00152]^t$$

Mode 1: Element 1 SE 45.2%

          Element 2 SE 54.8%

          Element 1 KE 16.5%

          Element 2 KE 83.5%

Mode 2: Element 1 SE 55.5%

          Element 2 SE 44.5%

          Element 1 KE 83.5%

          Element 2 KE 16.5%

Mode 3: Element 1 SE 99.3%
Element 2 SE 0.7%
Element 1 KE 100%
Element 2 KE 0%

Exercise 8: *Repeat exercise 6 with the beam modeled by two finite elements; see Figure* 9.

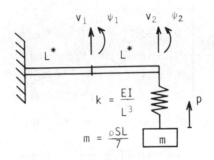

Figure 9

To keep the same matrix expressions, lct $L^* = L/2$. Then

$$T = \frac{1}{2}\frac{\rho SL^*}{420}\begin{bmatrix} v_1^{\circ} \\ \psi_1^{\circ} \end{bmatrix}^t \begin{bmatrix} 156 & -22L^* \\ -22L^* & 4L^{*2} \end{bmatrix}\begin{bmatrix} v_1^{\circ} \\ \psi_1^{\circ} \end{bmatrix}$$

$$+ \frac{1}{2}\frac{\rho SL^*}{420}\begin{bmatrix} v_1^{\circ} \\ \psi_1^{\circ} \\ v_2^{\circ} \\ \psi_2^{\circ} \end{bmatrix}^t \begin{bmatrix} 156 & 22L^* & 54 & -13L^* \\ 22L^* & 4L^{*2} & 13L^* & -3L^{*2} \\ 54 & 13L^* & 156 & -22L^* \\ -13L^* & -3L^{*2} & -22L^* & 4L^{*2} \end{bmatrix}\begin{bmatrix} v_1^{\circ} \\ \psi_1^{\circ} \\ v_2^{\circ} \\ \psi_2^{\circ} \end{bmatrix}$$

$$+ \frac{1}{2}\frac{2\rho SL^*}{7}p^{\circ 2}$$

$$U = \frac{1}{2}\frac{EI}{L^{*3}}\begin{bmatrix} v_1 \\ \psi_1 \end{bmatrix}^t \begin{bmatrix} 12 & -6L^* \\ -6L^* & 4L^{*2} \end{bmatrix}\begin{bmatrix} v_1 \\ \psi_1 \end{bmatrix}$$

$$+ \frac{1}{2}\frac{EI}{L^{*3}}\begin{bmatrix} v_1 \\ \psi_1 \\ v_2 \\ \psi_2 \end{bmatrix}\begin{bmatrix} 12 & 6L^* & -12 & 6L^* \\ 6L^* & 4L^{*2} & -6L^* & 2L^{*2} \\ -12 & -6L^* & 12 & -6L^* \\ 6L^* & 2L^{*2} & -6L^* & 4L^{*2} \end{bmatrix}\begin{bmatrix} v_1 \\ \psi_1 \\ v_2 \\ \psi_2 \end{bmatrix}$$

$$+ \frac{1}{2}\frac{EI}{8L^{*3}}(v_2 - p)^2$$

and applying Lagrange's equations gives

$$\rho SL^*\begin{bmatrix} 312/420 & 0 & 54/420 & -13L^*/420 & 0 \\ 0 & 8L^{*2}/420 & 13L^*/420 & -3L^{*2}/420 & 0 \\ 54/420 & 13L^*/420 & 156/420 & -22L^*/420 & 0 \\ -13L^*/420 & -3L^{*2}/420 & -22L^*/420 & 4L^{*2}/420 & 0 \\ 0 & 0 & 0 & 0 & 2/7 \end{bmatrix}\begin{bmatrix} v_1^{\infty} \\ \psi_1^{\infty} \\ v_2^{\infty} \\ \psi_2^{\infty} \\ p^{\infty} \end{bmatrix}$$

$$+\frac{EI}{L^{*3}}\begin{bmatrix} 24 & 0 & -12 & 6L^* & 0 \\ 0 & 8L^{*2} & -6L^* & 2L^{*2} & 0 \\ -12 & -6L^* & 12.125 & -6L^* & -0.125 \\ 6L^* & 2L^{*2} & -6L^* & 4L^{*2} & 0 \\ 0 & 0 & -0.125 & 0 & 0.125 \end{bmatrix}\begin{bmatrix} v_1 \\ \psi_1 \\ v_2 \\ \psi_2 \\ p \end{bmatrix}=0$$

from which one obtains with the aid of program no. 5:

$$\omega_1 = \frac{2.143}{L^2}\sqrt{\frac{EI}{\rho S}}; \qquad \phi_1 = \left[0.1108, \frac{0.3918}{L}, 0.3440, \frac{0.5008}{L}, 1\right]^t$$

$$\omega_2 = \frac{4.323}{L^2}\sqrt{\frac{EI}{\rho S}}; \qquad \phi_2 = \left[0.3545, \frac{1.184}{L}, 1, \frac{1.309}{L}, -0.5991\right]^t$$

$$\omega_3 = \frac{22.32}{L^2}\sqrt{\frac{EI}{\rho S}}; \qquad \phi_3 = \left[-0.2990, \frac{0.1872}{L}, 0.4187, \frac{2}{L}, -0.0060\right]^t$$

Exercise 9: Consider the system of exercise 7. Calculate the first three frequencies and their modes using computer program no. 11 and with three elements for the beam.

$\omega_1 = 27.13$ rad/sec; $\qquad \phi_1 = [0.0531, 0.1478, 0.1819, 0.2279,$
$$0.3441, 0.2504, 1]^t$$

$\omega_2 = 54.70$ rad/sec; $\qquad \phi_2 = [0.1752, 0.4719, 0.5626, 0.6475, 1,$
$$0.6544, -0.6000]^t$$

$\omega_3 = 281.0$ rad/sec; $\qquad \phi_3 = [-0.2466, -0.3672, -0.1749, 0.6223,$
$$0.4213, 1, -0.0061]^t$$

Exercise 10: Consider a clamped–clamped beam. Calculate the six lowest frequencies using first the exact theory and then a finite element model consisting of four elements (see Figure 10). Use the consistent mass matrix. In the latter case, use program no. 11. Use SI units with $L = 1\,m$; $E = 2 \times 10^{11}\,N/m^2$; $\rho = 7800\,kg/m^3$; $I = 10^{-6}\,m^4$; $S = 10^{-2}\,m^2$.

The results for ω in rad/sec are as follows:

Exact	70.81	195.2	382.6	632.5	944.9	1320
FEM	70.90	197.0	390.8	739.4	1223	1970

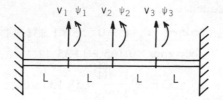

Figure 10

Exercise 11: *Consider the same system as in exercise 4. Add a mass of $\rho SL/10$ at the free end. Using program no. 5, calculate the values of the two lowest frequencies ω_1^*, ω_2^*. Then considering the added mass to be a small modification of the structure, calculate the values ω_1^{**}, ω_2^{**}.*

In the mass matrix of exercise 4, it is necessary to increase m_{55} by $\rho SL/10$; then:

$$\omega_1^* = \frac{3.302}{(3L)^2} \sqrt{\frac{EI}{\rho S}}$$

$$\omega_2^* = \frac{20.88}{(3L)^2} \sqrt{\frac{EI}{\rho S}}$$

and from the structural modification method of section 5.3:

$$\omega_1^{**} = \frac{3.303}{(3L)^2} \sqrt{\frac{EI}{\rho S}}$$

$$\omega_2^{**} = \frac{20.75}{(3L)^2} \sqrt{\frac{EI}{\rho S}}$$

Exercise 12: *Consider the same system as in exercise 6. However, in this case the thickness of the beam h is multiplied by 1.1. Using the results of exercise 7, find the influence of this structural modification on the first three frequencies. Compare these results to the results obtained using program no. 11.*

The strain energy of a beam in bending is proportional to h^3 and the kinetic energy to h. Hence for the first frequency

$$U_1 \text{ becomes } \quad 0.548 + 0.452 \times 1.331 = 1.150$$

$$T_1 \text{ becomes } \quad 0.835 + 0.165 \times 1.1 = 1.017$$

from which

$$\omega_1^* = 27.13 \sqrt{\frac{1.150}{1.017}} = 28.85 \text{ rad/sec}$$

For the second frequency

$$U_2 \text{ becomes} \quad 0.445 + 0.555 \times 1.331 = 1.184$$

$$T_2 \text{ becomes} \quad 0.165 + 0.835 \times 1.1 = 1.084$$

$$\omega_2^* = 55.62 \sqrt{\frac{1.184}{1.084}} = 58.13 \text{ rad/sec}$$

For the third frequency

$$U_3 \text{ becomes} \quad 0.007 + 0.993 \times 1.331 = 1.329$$

$$T_3 \text{ becomes} \quad 0 \quad + 1.0 \quad \times 1.1 \quad = 1.1$$

$$\omega_3^* = 442.1 \sqrt{\frac{1.329}{1.1}} = 485.9 \text{ rad/sec}$$

The results obtained using program no. 11 are

$$\omega_1^{**} = 28.56 \text{ rad/sec} \qquad \omega_2^{**} = 57.55 \text{ rad/sec} \qquad \omega_3^{**} = 485.9 \text{ rad/sec}$$

The result $\omega_3^* = \omega_3^{**}$ is to be expected because in this mode the mass has no kinetic energy and the spring has a very low strain energy.

Exercise 13: *The plane frame shown in Figure* 11 *is clamped at* A *and* C *and has a rigid joint at* B. *With the notation of Figure* 12, *express* u_2, v_2, ψ_2 *at*

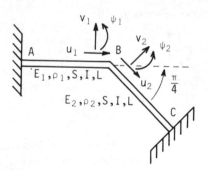

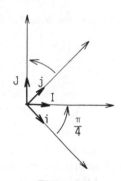

Figure 11   Figure 12

B *as functions of* u_1, v_1, ψ_1 *at* B *and obtain the kinetic energy* T *and strain energy* U *in terms of these latter variables. Use the consistent mass matrix and SI units with* $L = 1$ m, $S = 10^{-2}$ m², $I = 10^{-6}$ m⁴, $E_1 = 2 \times 10^{11}$ N/m², $\rho_1 = 7800$ kg/m³, $E_2 = 7 \times 10^{10}$ N/m², $\rho_2 = 2700$ kg/m.

$$u_2 \vec{i} + v_2 \vec{j} = u_1 \vec{I} + v_1 \vec{J}$$

from which

$$u_2 = u_1 \vec{I} \cdot \vec{i} + v_1 \vec{J} \cdot \vec{i}$$

$$v_2 = u_1 \vec{I} \cdot \vec{j} + v_1 \vec{J} \cdot \vec{j}$$

Then

$$\begin{bmatrix} u_2 \\ v_2 \\ \psi_2 \end{bmatrix} = \begin{bmatrix} \sqrt{2}/2 & -\sqrt{2}/2 & 0 \\ \sqrt{2}/2 & \sqrt{2}/2 & 0 \\ 0 & 0 & 1 \end{bmatrix} \begin{bmatrix} u_1 \\ v_1 \\ \psi_1 \end{bmatrix}$$

Using a consistent mass matrix,

$$T(1) = \frac{1}{2} \begin{bmatrix} u_1^{\circ} \\ v_1^{\circ} \\ \psi_1^{\circ} \end{bmatrix}^t \begin{bmatrix} 26 & 0 & 0 \\ 0 & 28.97 & -4.086 \\ 0 & -4.086 & 0.7429 \end{bmatrix} \begin{bmatrix} u_1^{\circ} \\ v_1^{\circ} \\ \psi_1^{\circ} \end{bmatrix}$$

$$T(2) = \frac{1}{2} \begin{bmatrix} u_1^{\circ} \\ v_1^{\circ} \\ \psi_1^{\circ} \end{bmatrix}^t \begin{bmatrix} \sqrt{2}/2 & -\sqrt{2}/2 & 0 \\ \sqrt{2}/2 & \sqrt{2}/2 & 0 \\ 0 & 0 & 1 \end{bmatrix}^t \begin{bmatrix} 9 & 0 & 0 \\ 0 & 10.03 & 1.414 \\ 0 & 1.414 & 0.2571 \end{bmatrix}$$

$$\times \begin{bmatrix} \sqrt{2}/2 & -\sqrt{2}/2 & 0 \\ \sqrt{2}/2 & \sqrt{2}/2 & 0 \\ 0 & 0 & 1 \end{bmatrix} \begin{bmatrix} u_1^{\circ} \\ v_1^{\circ} \\ \psi_1^{\circ} \end{bmatrix}$$

$$= \frac{1}{2} \begin{bmatrix} u_1^{\circ} \\ v_1^{\circ} \\ \psi_1^{\circ} \end{bmatrix}^t \begin{bmatrix} 9.514 & 0.5143 & 1 \\ 0.5143 & 9.514 & 1 \\ 0 & 1 & 0.2571 \end{bmatrix} \begin{bmatrix} u_1^{\circ} \\ v_1^{\circ} \\ \psi_1^{\circ} \end{bmatrix}$$

from which

$$T = \frac{1}{2} \begin{bmatrix} u_1^{\circ} \\ v_1^{\circ} \\ \psi_1^{\circ} \end{bmatrix}^t \begin{bmatrix} 35.51 & 0.5143 & 1 \\ 0.5143 & 38.49 & -3.086 \\ 1 & -3.086 & 1 \end{bmatrix} \begin{bmatrix} u_1^{\circ} \\ v_1^{\circ} \\ \psi_1^{\circ} \end{bmatrix}$$

Similarly,

$$U(1) = \frac{1}{2} \begin{bmatrix} u_1 \\ v_1 \\ \psi_1 \end{bmatrix}^t \begin{bmatrix} 2 \times 10^9 & 0 & 0 \\ 0 & 2.4 \times 10^6 & -1.2 \times 10^6 \\ 0 & -1.2 \times 10^6 & 0.8 \times 10^6 \end{bmatrix} \begin{bmatrix} u_1 \\ v_1 \\ \psi_1 \end{bmatrix}$$

$$U(2) = \frac{1}{2} \begin{bmatrix} u_1 \\ v_1 \\ \psi_1 \end{bmatrix}^t \begin{bmatrix} \sqrt{2}/2 & -\sqrt{2}/2 & 0 \\ \sqrt{2}/2 & \sqrt{2}/2 & 0 \\ 0 & 0 & 1 \end{bmatrix}^t \begin{bmatrix} 7 \times 10^8 & 0 & 0 \\ 0 & 8.4 \times 10^5 & 4.2 \times 10^5 \\ 0 & 4.2 \times 10^5 & 2.8 \times 10^5 \end{bmatrix}$$

$$\times \begin{bmatrix} \sqrt{2}/2 & -\sqrt{2}/2 & 0 \\ \sqrt{2}/2 & \sqrt{2}/2 & 0 \\ 0 & 0 & 1 \end{bmatrix} \begin{bmatrix} u_1 \\ v_1 \\ \psi_1 \end{bmatrix}$$

$$= \frac{1}{2} \begin{bmatrix} u_1 \\ v_1 \\ \psi_1 \end{bmatrix}^t \begin{bmatrix} 3.504 \times 10^8 & -3.496 \times 10^8 & 2.97 \times 10^5 \\ -3.496 \times 10^8 & 3.504 \times 10^8 & 2.97 \times 10^5 \\ 2.97 \times 10^5 & 2.97 \times 10^5 & 2.8 \times 10^5 \end{bmatrix} \begin{bmatrix} u_1 \\ v_1 \\ \psi_1 \end{bmatrix}$$

and

$$U = \frac{1}{2} \begin{bmatrix} u_1 \\ v_1 \\ \psi_1 \end{bmatrix}^t \begin{bmatrix} 2.35 \times 10^9 & -3.496 \times 10^8 & 2.97 \times 10^5 \\ -3.496 \times 10^8 & 3.528 \times 10^8 & -9.03 \times 10^5 \\ 2.97 \times 10^5 & -9.03 \times 10^5 & 1.08 \times 10^6 \end{bmatrix} \begin{bmatrix} u_1 \\ v_1 \\ \psi_1 \end{bmatrix}$$

Using computer program no. 5, one finds

$$\omega_1 = 1030 \text{ rad/sec}; \qquad \phi_1 = [0.0010, -0.0087, 1]^t$$
$$\omega_2 = 3131 \text{ rad/sec}; \qquad \phi_2 = [0.0609, 0.3170, 1]^t$$
$$\omega_3 = 8619 \text{ rad/sec}; \qquad \phi_3 = [-0.4794, 0.1653, 1]^t$$

These results can be obtained directly using computer program no. 12.

Exercise 14: *Consider the clamped–clamped beam of exercise* 10. *Calculate the two lowest frequencies by using the method of substructures which employs free modes; see Figure* 13. *Use only the first mode of SS1, but retain all the degrees of freedom of SS2. Use program no. 5 to calculate frequencies and modes.*

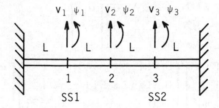

Figure 13

It is necessary to find the first mode of SS1 with node 2 free; see

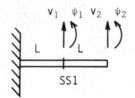

Figure 14

Figure 14. The consistent mass matrix and stiffness matrix are

$$M = \begin{bmatrix} 57.94 & 0 & 10.03 & -2.414 \\ 0 & 1.486 & 2.414 & -0.5571 \\ 10.03 & 2.414 & 28.97 & -4.086 \\ -0.2414 & -0.5571 & -4.086 & 0.7429 \end{bmatrix}$$

and

$$K = 2 \times 10^5 \begin{bmatrix} 24 & 0 & -12 & 6 \\ 0 & 8 & -6 & 2 \\ -12 & -6 & 12 & -6 \\ 6 & 2 & -6 & 4 \end{bmatrix}$$

Computer program no. 5 gives

$$\omega_1 = 44.53 \text{ rad/sec}; \qquad \phi_1 = \begin{bmatrix} 0.3395 \\ 0.5815 \\ 1 \\ 0.6882 \end{bmatrix}$$

and permits the change of variable

$$\begin{bmatrix} v_1 \\ \psi_1 \end{bmatrix} = \begin{bmatrix} 0.3395 \\ 0.5815 \end{bmatrix} q_1$$

Using the variables q_1, v_2, ψ_2 and completing all calculations gives

$$T_{SS1} = \frac{1}{2} \begin{bmatrix} q_1^\circ \\ v_2^\circ \\ \psi_2^\circ \end{bmatrix}^t \begin{bmatrix} 7.182 & 4.809 & -1.144 \\ 4.809 & 28.97 & -4.086 \\ -1.144 & -4.086 & 0.7429 \end{bmatrix} \begin{bmatrix} q_1^\circ \\ v_2^\circ \\ \psi_2^\circ \end{bmatrix}$$

$$U_{SS1} = \frac{10^5}{2} \begin{bmatrix} q_1 \\ v_2 \\ \psi_2 \end{bmatrix}^t \begin{bmatrix} 10.94 & -15.13 & 6.4 \\ -15.13 & 24 & -12 \\ 6.4 & -12 & 8 \end{bmatrix} \begin{bmatrix} q_1 \\ v_2 \\ \psi_2 \end{bmatrix}$$

For SS2 one obtains

$$T_{SS2} = \frac{1}{2} \begin{bmatrix} v_2^\circ \\ \psi_2^\circ \\ v_3^\circ \\ \psi_3^\circ \end{bmatrix}^t \begin{bmatrix} 28.97 & 4.086 & 10.03 & -2.414 \\ 4.086 & 0.7429 & 2.414 & -0.5571 \\ 10.03 & 2.414 & 57.94 & 0 \\ -2.414 & -0.5571 & 0 & 1.486 \end{bmatrix} \begin{bmatrix} v_2^\circ \\ \psi_2^\circ \\ v_3^\circ \\ \psi_3^\circ \end{bmatrix}$$

$$U_{SS2} = \frac{10^5}{2} \begin{bmatrix} v_2 \\ \psi_2 \\ v_3 \\ \psi_3 \end{bmatrix}^t \begin{bmatrix} 24 & 12 & -24 & 12 \\ 12 & 8 & -12 & 4 \\ -24 & -12 & 48 & 0 \\ 12 & 4 & 0 & 16 \end{bmatrix} \begin{bmatrix} v_2 \\ \psi_2 \\ v_3 \\ \psi_3 \end{bmatrix}$$

The mass and stiffness matrices for the total structure can now be assembled.

$$M = \begin{bmatrix} 7.182 & 4.809 & -1.144 & 0 & 0 \\ 4.809 & 57.94 & 0 & 10.03 & -2.414 \\ -1.144 & 0 & 1.486 & 2.414 & -0.5571 \\ 0 & 10.03 & 2.414 & 57.94 & 0 \\ 0 & -2.414 & -0.5571 & 0 & 1.486 \end{bmatrix}$$

$$K = 10^5 \begin{bmatrix} 10.94 & -15.13 & 6.4 & 0 & 0 \\ -15.13 & 48 & 0 & -24 & 12 \\ 6.4 & 0 & 16 & -12 & 4 \\ 0 & -24 & -12 & 48 & 0 \\ 0 & 12 & 4 & 0 & 16 \end{bmatrix}$$

Program no. 5 gives

$$\omega_1 = 72.16 \text{ rad/sec}; \qquad \phi_1 = [1, 0.6981, 0.0255, 0.3874, -0.5381]^t$$
$$\omega_2 = 216.4 \text{ rad/sec}; \qquad \phi_2 = [1, 0.0754, -0.9043, -0.4652, -0.1871]^t$$

In order to recover the modes in terms of the initial variables, one uses the relation among v_1, ψ_1, and q_1. For example, for the lowest frequency one finds the following mode in terms of the degrees of freedom v_1, ψ_1, v_2, ψ_2, v_3, ψ_3:

$$[0.4864, 0.8330, 1, 0.0365, 0.5550, -0.7708]^t$$

where the mode has been normalized so that its largest component is unity.

Exercise 15: *Consider the same system as in exercise* 14 *but use constrained modes for* SS1.

It is necessary to find the first mode of SS1 with node 2 clamped; see

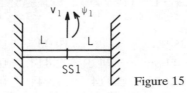

Figure 15

Figure 15. The consistent mass matrix and stiffness matrix are

$$M = \begin{bmatrix} 57.94 & 0 \\ 0 & 1.486 \end{bmatrix}; \qquad k = 10^5 \begin{bmatrix} 48 & 0 \\ 0 & 16 \end{bmatrix}$$

from which one finds

$$\omega_1 = 287.8 \text{ rad/sec}; \qquad \phi_1 = \begin{bmatrix} 1 \\ 0 \end{bmatrix}$$

The static displacement is found from

$$2 \times 10^5 \begin{bmatrix} 24 & 0 & -12 & 6 \\ 0 & 8 & -6 & 2 \\ -12 & -6 & 12 & -6 \\ 6 & 2 & -6 & 4 \end{bmatrix} \begin{bmatrix} v_1 \\ \psi_1 \\ v_2 \\ \psi_2 \end{bmatrix} = 0$$

From the first two lines one obtains

$$\begin{bmatrix} v_1 \\ \psi_1 \end{bmatrix} = \begin{bmatrix} 0.5 & -0.25 \\ 0.75 & -0.25 \end{bmatrix} \begin{bmatrix} v_2 \\ \psi_2 \end{bmatrix}$$

The transformation matrix becomes

$$
\begin{bmatrix} v_1 \\ \psi_1 \\ v_2 \\ \psi_2 \end{bmatrix} = \begin{bmatrix} 1 & 0.5 & -0.25 \\ 0 & 0.75 & -0.25 \\ 0 & 1 & 0 \\ 0 & 0 & 1 \end{bmatrix} \begin{bmatrix} q_1 \\ v_2 \\ \psi_2 \end{bmatrix}
$$

and this leads to

$$
T_{SS1} = \frac{1}{2} \begin{bmatrix} q_1^\circ \\ v_2^\circ \\ \psi_2^\circ \end{bmatrix}^t \begin{bmatrix} 1 & 0.5 & -0.25 \\ 0 & 0.75 & -0.25 \\ 0 & 1 & 0 \\ 0 & 0 & 1 \end{bmatrix}^t
$$

$$
\times \begin{bmatrix} 57.94 & 0 & 10.03 & -2.414 \\ 0 & 1.486 & 2.414 & -0.5571 \\ 10.03 & 2.414 & 28.97 & -4.086 \\ -2.414 & -0.5571 & -4.086 & 0.7429 \end{bmatrix} \begin{bmatrix} 1 & 0.5 & -0.25 \\ 0 & 0.75 & -0.25 \\ 0 & 1 & 0 \\ 0 & 0 & 1 \end{bmatrix} \begin{bmatrix} q_1^\circ \\ v_2^\circ \\ \psi_2^\circ \end{bmatrix}
$$

$$
= \frac{1}{2} \begin{bmatrix} q_1^\circ \\ v_2^\circ \\ \psi_2^\circ \end{bmatrix}^t \begin{bmatrix} 57.94 & 39 & -16.90 \\ 39 & 57.94 & -16.34 \\ 16.90 & 16.34 & 5.943 \end{bmatrix} \begin{bmatrix} q_1^\circ \\ v_2^\circ \\ \psi_2^\circ \end{bmatrix}
$$

Proceeding in a similar manner:

$$
U_{SS1} = \frac{10^5}{2} \begin{bmatrix} q_1 \\ v_2 \\ \psi_2 \end{bmatrix}^t \begin{bmatrix} 48 & 0 & 0 \\ 0 & 3 & -3 \\ 0 & -3 & 4 \end{bmatrix} \begin{bmatrix} q_1 \\ v_2 \\ \psi_2 \end{bmatrix}
$$

Using the expression for T_{SS2} and U_{SS2} from exercise 14, the mass and stiffness matrices for the total structure can be assembled.

$$
M = \begin{bmatrix} 57.94 & 39 & -16.90 & 0 & 0 \\ 39 & 86.91 & -12.26 & 10.03 & -2.414 \\ -16.90 & -12.26 & 6.686 & 2.414 & -0.5571 \\ 0 & 10.03 & 2.414 & 57.94 & 0 \\ 0 & -2.414 & -0.5571 & 0 & 1.486 \end{bmatrix}
$$

$$
K = 10^5 \begin{bmatrix} 48 & 0 & 0 & 0 & 0 \\ 0 & 27 & 9 & -24 & 12 \\ 0 & 9 & 12 & -12 & 4 \\ 0 & -24 & -12 & 48 & 0 \\ 0 & 12 & 4 & 0 & 16 \end{bmatrix}
$$

Computer program no. 5 gives

$$\omega_1 = 70.91 \text{ rad/sec}; \qquad \phi_1 = [0.0435, 1, 0, 0.5435, -0.7612]^t$$
$$\omega_2 = 197.0 \text{ rad/sec}; \qquad \phi_2 = [-0.2576, 0, 1, 0.5067, -0.2729]^t$$

In order to recover the modes in terms of the initial variable, one uses the above transformation matrix. For example, for the lowest frequency one finds the following mode in terms of the degrees of freedom v_1, ψ_1, v_2, ψ_2, v_3, ψ_3:

$$[0.5435, 0.7500, 1, 0, 0.5435, -0.7612]^t$$

where the mode has been normalized so that its largest component is unity.

Exercise 16; *Consider the same system and same modeling procedure as in exercise 14. In this case use just the first mode of both* SS1 *and* SS2; *see Figure* 16.

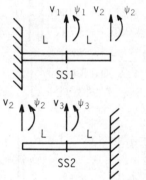

Figure 16

The appropriate expressions for T_{SS1} and U_{SS1} are given in exercise 14. The first frequency and mode of SS2 are

$$\omega_1 = 44.53 \text{ rad/sec}; \qquad \phi_1 = \begin{bmatrix} 1 \\ -0.6882 \\ 0.3395 \\ -0.5815 \end{bmatrix}$$

which permits the change of variable:

$$\begin{bmatrix} v_3 \\ \psi_3 \end{bmatrix} = \begin{bmatrix} 0.3395 \\ -0.5815 \end{bmatrix} q_2$$

Using this in the expressions for T_{SS2} and U_{SS2} gives

$$T_{\text{SS2}} = \frac{1}{2} \begin{bmatrix} v_2^\circ \\ \psi_2^\circ \\ q_2^\circ \end{bmatrix}^t \begin{bmatrix} 28.97 & 4.086 & 4.809 \\ 4.086 & 0.7429 & 1.144 \\ 4.809 & 1.144 & 7.182 \end{bmatrix} \begin{bmatrix} v_2^\circ \\ \psi_2^\circ \\ q_2^\circ \end{bmatrix}$$

$$U_{ss2} = \frac{10^5}{2} \begin{bmatrix} v_2 \\ \psi_2 \\ q_2 \end{bmatrix}^t \begin{bmatrix} 24 & 12 & -15.13 \\ 12 & 8 & -6.4 \\ -15.13 & -6.4 & 10.94 \end{bmatrix} \begin{bmatrix} v_2 \\ \psi_2 \\ q_2 \end{bmatrix}$$

The mass and stiffness matrices for the total structure can then be assembled.

$$M = \begin{bmatrix} 7.182 & 4.809 & -1.144 & 0 \\ 4.809 & 57.94 & 0 & 4.809 \\ -1.144 & 0 & 1.486 & 1.144 \\ 0 & 4.809 & 1.144 & 7.182 \end{bmatrix}$$

$$K = 10^5 \begin{bmatrix} 10.94 & -15.13 & 6.4 & 0 \\ -15.13 & 48 & 0 & -15.13 \\ 6.4 & 0 & 16 & -6.4 \\ 0 & -15.13 & -6.4 & 10.94 \end{bmatrix}$$

where the variables are, in order, q_1, v_2, ψ_2, q_2. Computer program no. 5 gives

$$\omega_1 = 73.58 \text{ rad/sec}; \qquad \phi_1 = [1, 0.6860, 0, 1]^t$$
$$\omega_2 = 244.7 \text{ rad/sec}; \qquad \phi_2 = [1, 0, -0.9378, -1]^t$$

Using the transformations between v_1, ψ_1, and q_1 and v_2, ψ_2, and q_2, the mode associated with the lowest frequency in terms of the degrees of freedom $v_1, \psi_1, v_2, \psi_2, v_3, \psi_3$ is

$$[0.4950, 0.8477, 1, 0, 0.4950, 0.8477]^t$$

where the mode has been normalized so that its largest component is unity.

6
Experimental Methods

This chapter is addressed to readers who are not expert in the measurement of mechanical vibrations. It is intended primarily as an introduction to current equipment and techniques but it should also enable the reader to perform simple tests and to understand more advanced treatments.

The experimental study of a structure is necessary for a number of reasons: to determine its performance under actual operating conditions; to establish the accuracy of analytical predictions; or to obtain its dynamic characteristics when calculation is too uncertain or too expensive. This last case can be encountered when the structure is so complex that it is not amenable to economical calculation by the computer which is available or when the values of damping or the boundary conditions are inadequately known.

Among the most common measurements are the vibration response due to a sinusoidal excitation which is swept over a wide range of frequencies, the vibration response due to broadband noise, or the vibration response due to impact. These last two types of measurements are becoming more frequent because they are closely related to modern techniques of signal analysis and to recent developments in equipment for real-time signal display.

In what follows, common kinds of vibration transducers and exciters are presented as well as a brief introduction to the measurement of the dynamic characteristics of a structure subjected to a known excitation or under operating conditions.

The contents of the chapter are as follows:
6.1   Transducers
6.2   Exciters
6.3   Measurements.

6.1   Transducers

The basic principles of the most common vibration transducers are presented without a discussion of the technical details of their construction. Vibration transducers fall into one of two categories – contact or noncontact – and each is discussed in turn.

6.1.1 Transducers in contact with the structure

Piezoelectric transducers

These transducers use a property of piezoelectric quartz or ceramic materials which have asymmetric internal charge distributions. If a force is applied to such materials, the lattice is deformed and equal external charges of opposite polarity appear on the opposite sides of the crystal. To permit measurement, this charge, which is proportional to the applied force *F*, must be converted into a voltage by one of the following devices:

1. A charge-sensitive preamplifier. This device converts the charge into a voltage in such a way that the output voltage is essentially independent of the length of cable between the transducer and the preamplifier.

2. A voltage-sensitive preamplifier. The output of this device is directly proportional to the voltage difference across the transducer terminals. However, the signal is affected by cable length and often a correction for this effect must be applied.

Transducers which incorporate the sensor and the signal preamplifier within a single unit have been available for several years. The most common types of piezoelectric transducers are accelerometers and force gages.

Accelerometer. A slice of piezoelectric quartz crystal Q is attached to a mass M on one face and to a rigid base B on the other. The resulting transducer is attached to the structure by its base, as shown in Figure 1.

As a first approximation, this transducer is modeled as a mass *m* in series with a spring of stiffness *k*; see Figure 2. Such devices are given the generic name of seismic pickups.

The displacements x_1 and x_2 are related by the differential equation of motion for mass *m*, i.e.

$$mx_2^{\infty} + k(x_2 - x_1) = 0 \qquad (1)$$

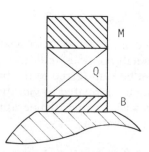

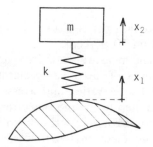

Figure 1   Schematic of an accelerometer

Figure 2   Single degree-of-freedom model of the accelerometer in Figure 1

and the force acting on the quartz crystal is

$$f = k(x_1 - x_2) \tag{2}$$

Let $x_1 = X_1 \sin \Omega t$, then in the steady-state motion

$$x_2 = X_2 \sin \Omega t \tag{3}$$

and (1) becomes

$$-m\Omega^2 X_2 + k(X_2 - X_1) = 0 \tag{4}$$

from which

$$X_2 = \frac{X_1}{1 - (\Omega/\omega)^2} \tag{5}$$

where

$$\omega^2 = \frac{k}{m}$$

Also, (2) becomes

$$f = -kX_1 \frac{(\Omega/\omega)^2}{1 - (\Omega/\omega)^2} \sin \Omega t = F \sin \Omega t \tag{6}$$

For $\Omega \ll \omega$, (6) becomes

$$f \approx -k\left(\frac{\Omega}{\omega}\right)^2 X_1 \approx -m\Omega^2 X_1 \tag{7}$$

and thus the force f and hence the surface electric charge produced is proportional to the acceleration of the base. The transducer therefore acts as an accelerometer.

Most accelerometers are supplied with a calibration chart similar to that shown in Figure 3. If s_f is the transducer sensitivity at an arbitrary frequency f and s_0 is the sensitivity at a low frequency, equations (6) and (7) give

$$\frac{s_f}{s_0} = \frac{1}{1 - (\Omega/\omega)^2} \tag{8}$$

Equation (8) is plotted in Figure 3 and shows that if $\Omega/\omega < 0.3$, $s_f = s_0$ to within 10 percent.

The measurements from an accelerometer can contain errors from several sources. Transverse acceleration will induce a small spurious indication of axial acceleration and stresses in the structure can be transferred to the base and then to the piezoelectric material to give a signal unrelated to vibration. Also, errors can arise from harsh environments such as high temperature, strong electromagnetic fields, or intense acoustic radiation.

Force transducer. A force transducer is constructed from piezoelectric material which is compressed between two disks of mass m_1 and m_2; see the

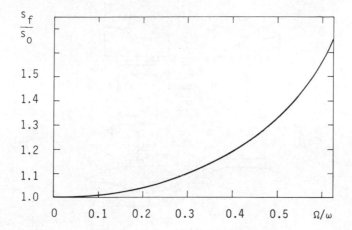

Figure 3   Calibration chart of an accelerometer

schematic representation in Figure 4. An axial force causes the production of surface charge on the faces of the piezoelectric material. This is transformed into a voltage difference via a charge-sensitive or voltage-sensitive preamplifier.

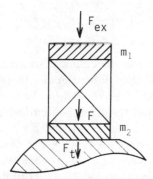

Figure 4   Schematic of a force transducer

The signal generated is not solely proportional to the force F_t transmitted to the structure. As a first approximation, it is proportional to the force F which is transmitted to mass m_2. The difference between F and F_t is equal to the inertia force of mass m_2. If the structure is small, it is necessary to compensate the output signal for the inertia force of m_2 in order to obtain an accurate measure of F_t. To cancel this mass effect, the corresponding inertia force is subtracted by calculation or with the aid of an analog circuit.

Like piezoelectric accelerometers, piezoelectric force transducers are sensitive to their conditions of use and to their environment. When they are subjected to base bending moments they will give measurement errors and can even be destroyed. These undesirable effects can be controlled by the simple means shown in Figure 5.

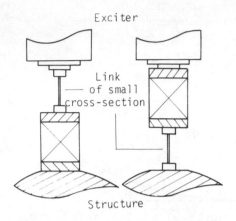

Figure 5 Mounting of a force transducer to minimize moment transfer

The mechanical link between the vibration exciter and the force transducer or between the force transducer and the structure is a beam of small cross-sectional dimension. Such a beam can have a high ratio of axial stiffness to bending stiffness and thereby limits the moment transferred to the transducer.

Electric transducers

Strain gage. This device is based on the fact that mechanical strain affects electrical properties such as the resistance R; these small changes in resistance can be easily measured with the aid of a Wheatstone bridge. The gage is bonded directly on the structure so that its strain is identical to the strain of the structure.

Strain gages are of two types. The most common type is the electrical resistance gage based on the variation of resistance of a wire with its elongation; the other type is based on the piezoresistive effect exhibited by certain semiconductor materials.

The relation between the percentage change in resistance and the longitudinal strain of a wire of length L is given by:

$$\frac{\Delta R}{R} = \alpha \frac{\Delta L}{L} \tag{9}$$

For the first type of gage, the so-called gage-factor α is in the neighborhood of 2 for most wire materials. This type of gage can measure static as well as dynamic strain. For the second type, the gage factor is of the order of several hundred and the measuring equipment can therefore be less sensitive; however, only dynamic strains can be measured. If either of these types is bonded to a metal structure, the surface strain can be measured without

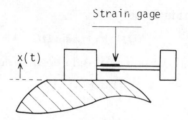

Figure 6 Strain gage vibration transducer

introducing significant perturbation to the structural behavior. If they are
made part of a transducer, they can be used to measure strains which are
proportional either to acceleration – a vibration transducer – or to the force
applied to the base – a force transducer. Figure 6 shows a strain gage being
used as a vibration transducer. Note that the gage is mounted near the root
of the cantilever where the strain is high.

6.1.2 Transducers not in contact with the structure

It is obvious that proximity pickups, i.e. noncontacting transducers, do not
affect the behavior of a structure while the preceding contact transducers
add local mass and stiffness. These local effects are important if the structure
is light or flexible. On the other hand, noncontacting transducers will
generate significant harmonics of the forcing frequency if the displacement
of the structure is large.

Capacitive transducer

The structure is at a DC potential V_S. The transducer, the plate A, located
at rest at a distance h_0 from the surface of the structure, is at a DC potential
V_A; see Figure 7. The potential difference $V = V_A - V_S$ is maintained
constant.

The transducer is assumed fixed in space. When the structure vibrates the
capacitance of the capacitor formed by the structure and the plate varies.
The variation in charge thus produced is transformed into a voltage $v(t)$ by a
charge-sensitive preamplifier. Let the time-varying distance between the

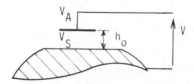

Figure 7 Capacitive transducer

structure and the transducer be

$$h(t) = h_0 + d \sin \Omega t \tag{10}$$

The charge $Q(t)$ and the potential difference V are related by

$$Q(t) = C(t) \cdot V \tag{11}$$

where V is a constant.

Reference to the formulas for the capacitance of plate capacitors shows that to first order

$$Q(t) = \frac{\alpha}{h(t)} \tag{12}$$

where α is a constant. The voltage output of a charge-sensitive preamplifier is then

$$v(t) = \frac{\beta}{h(t)} = \frac{\beta}{h_0 + d \sin \Omega t} \tag{13}$$

with $\beta = $ constant. If d is small compared to h_0, (13) becomes

$$v(t) \approx \frac{\beta}{h_0}\left(1 - \frac{d}{h_0} \sin \Omega t\right) \tag{14}$$

The alternating part of $v(t)$ is therefore proportional to the vibration displacement of the structure. If d is not small compared to h_0 and if one supposes that $v(t)$ is proportional to d in accordance with (14), one introduces a percentage error having a peak-to-peak value given by

$$\delta = \frac{(d/h_0)^2}{1 - (d/h_0)^2} \tag{15}$$

This is plotted in Figure 8.

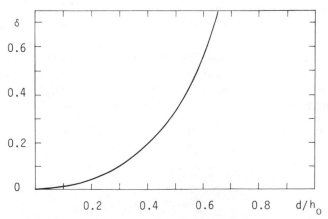

Figure 8 Peak-to-peak percentage error of a capacitive transducer

Magnetic transducer

A schematic diagram of a magnetic transducer is given in Figure 9. It shows a magnetic circuit composed of a coil wound on a permanent magnet and a structure made of a ferromagnetic material. If the structural material is nonmagnetic, such as aluminium or austenitic steel, a thin ferromagnetic disk can be bonded to the structure just beneath the transducer.

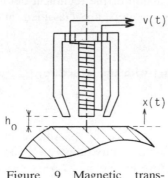

Figure 9 Magnetic transducer

At rest, the transducer and the structure are separated by a small distance h_0. The magnetic flux across h_0 is ϕ.

When the structure vibrates relative to the fixed transducer, the flux varies and a voltage is induced in the coil. This voltage, $v(t)$, is proportional to the product of the rate of change of ϕ with t and to the velocity of the structure. If the terminals of the electric circuit are connected to a high impedance and if the amplitude of sinusoidal displacement d is small compared to h_0, it can be shown that a first approximation is

$$v(t) = \alpha d\Omega \sin \Omega t \qquad (16)$$

The magnetic transducer is therefore a velocity pickup. Its sensitivity depends on h_0 but an increase in sensitivity is accompanied by an increase in nonlinearity.

Eddy-current transducer

This transducer consists of a coil in which flows a high-frequency current, usually in the megahertz range. The coil produces a magnetic field and if a structure made of a conductive material is placed in this field, there is a dissipation of energy which is a function of the material and the distance separating the transducer from the structure. This dissipation of energy affects the inductance of the coil and this is reflected in the amplitude variation of the high-frequency voltage across the coil. After demodulation, the coil voltage is proportional, under certain conditions, to the distance between the transducer and the structure. The transducer therefore acts as a displacement pickup. This transducer can be used for static as well as dynamic measurements. Its sensitivity is function of the nature of material.

6.2 Exciters

As was the case for transducers, the basic principles of vibration exciters are presented without a discussion of the technical details of their construction. Also as for transducers, exciters are classified into contacting and non-contacting types.

6.2.1 Exciters in contact with the structure

Electrodynamic shaker

This shaker generates a force by passing a current I through a coil mounted in a magnetic field of flux density B; see Figure 10. The force is transmitted to the structure by means of a fixture. The force generated in the exciter coil is

$$F = 2\pi RNBI \tag{17}$$

where N is the number of turns in the coil and R is the coil radius. If the current variation is sinusoidal of frequency Ω, equation (17) becomes

$$F = 2\pi RNBI_0 \sin \Omega t \tag{18}$$

Since the exciter is attached to the structure, it can modify the dynamic behavior of the structure by introducing supplementary springs and masses. Electrodynamic exciters are capable of generating a wide variety of force shapes; in particular, harmonic forces can be generated in a very pure form. The useful operating range of frequency is usually from a few hertz to about

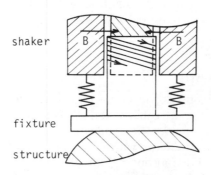

Figure 10   Electrodynamic shaker
mounted to a structure by means
of a fixture

10 kHz. The coil displacements are usually small, the maximum being of the order of a few centimeters, peak-to-peak.

Electrohydraulic shaker

This consists of a hydraulic cylinder and piston together with a servo-control system. The electronic control system allows the vibration signal to be generated easily and the hydraulic drive system can be designed to give large displacements and high forces. The useful operating frequency range is usually from fractions of a hertz to a few hundred hertz.

6.2.2 Exciters not in contact with the structure

These have the advantage of not affecting the dynamic behavior of the structure.

Magnetic exciter

This is essentially a magnetic transducer used as a driver rather than as a pickup. It is a convenient way of delivering low force levels with a device of simple design. Usually, the coil is wound on a permanent magnet whose magnetic field is much stronger than that due to the current from the power amplifier; the permanent magnet is sometimes replaced by a DC electro-magnet. If the amplitude of motion of the structure is much less than the static separation distance, the excitation force has the form

$$F = F_0 + F_1 \cos \Omega t \tag{19}$$

where F_0 is the constant force of attraction.

In the case when the coil has a core of ferromagnetic material rather than a magnetic core, it can be shown that

$$F = F_0 + F_1 \cos 2\Omega t \tag{20}$$

If the amplitude of motion is large, the excitation force contains non-negligible harmonics and it is necessary to use a tracking filter.

When the structure is nonmagnetic, direct excitation can be obtained by fixing a thin disk of ferromagnetic material to the surface of the structure.

6.3 Measurements

Under laboratory conditions, the experimenter can impose a known excitation on a structure in order to determine its dynamic characteristics; that is, its resonant frequencies, modes, and modal damping. The imposed excitation can be a sinusoidal frequency sweep, a white noise, or an impact.

Under field conditions, the experimenter cannot always impose an excitation of his choice. It is necessary to measure and analyze the operating behavior of the structure in order to detect problems and assess performance.

6.3.1 Measurement system for sinusoïdal excitation

If the exciter is in contact with the structure, one can eliminate the influence of the fixture and exciter mass and stiffness by using transducer 1, depicted in Figure 11, as a reference. If this transducer is a force transducer, the measurement is the response to an imposed force; if it is a displacement transducer, the response is due to imposed displacement.

If transducers 1 and 2 measure the same quantities, either forces or displacements, the ratio of these like quantities is the force or displacement transmissibility. If transducers 1 and 2 measure different quantities, the ratios of these quantities can, be x_2^{∞}/F_1, acceleration mobility, x_2°/F_1, velocity mobility, or x_2/F_1, displacement mobility. The latter quantity is sometimes called the dynamic compliance or the receptance. If the reciprocals of these quantities are measured, i.e., F_1/x_2^{∞}, F_1/x_2°, F_1/x_2, one obtains, respectively, the dynamic mass, the mechanical impedance, or the dynamic stiffness. If points 1 and 2 coincide, these ratios are driving point quantities; if points 1 and 2 are different, they are transfer quantities.

As shown in Figure 11, an oscilloscope is used to monitor the quality of the electrical signals. The reasons for this are:

(a) the sinusoïdal excitation may not be sufficiently pure

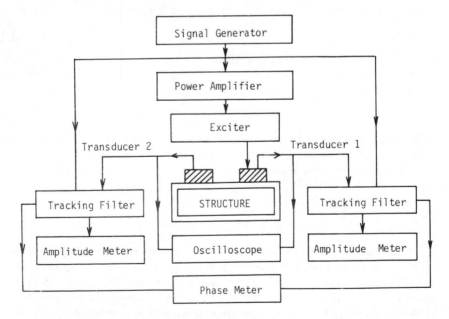

Figure 11   Measurement system for sinusoidal excitation

(b) the signal level can be reduced if it is too large or increased if it is below the noise level

(c) system nonlinearities can be detected.

The experimenter is thereby made aware when it is necessary to adjust the excitation level or to be particularly careful in the processing of the response. The tracking filter guarantees that the response is measured only at the excitation frequency. The combination of a shaker which can deliver a very pure sine wave and a tracking filter to minimize harmonics results in a response with very little extraneous noise. This is a distinct advantage of the measuring system. The transducer signal amplitudes and phase difference are the most important quantities to be measured.

Determination of dynamic characteristics

The first step is to plot the modulus and phase of the response over the full frequency range of interest. Such a frequency response plot allows estimation of the dynamic characteristics of the structure and thereby the amount of damping and position of resonant frequencies. If the structure is lightly damped and the modes well separated, each peak of the modulus plot corresponds to a resonant frequency. The corresponding mode can then be determined by dwelling at each resonant frequency and making a spatial survey of the structure. Lastly, a narrow sweep can be made around each resonant frequency to establish the half-power bandwidth from which the modal damping or Q-factor can be calculated.

If the resonant frequencies are strongly coupled, either because of large damping or close spacing, it is difficult to extract the individual modal characteristics of the structure accurately. In these circumstances, specialized computer techniques must be used, an explanation of which is beyond the scope of this text.

A polar plot of the modulus and phase of the response is known as a Nyquist diagram. This diagram is better suited for extracting modal information because the effects of frequency, damping, stiffness, and mass are magnified. If the damping is of the structural type, the response variable in the Nyquist diagram is usually displacement mobility (see exercise 16 of Chapter 1), and if the damping is of the viscous type, the response variable is usually velocity mobility.

As mentioned previously, the determination of the mode shape is done by exciting the structure at one of its resonant frequencies and surveying the resulting distribution of displacement amplitudes. Of particular interest are the nodal points or lines associated with each mode shape. A nodal line is a locus of points which has zero displacement amplitude. Nodal lines can be identified by a number of methods which depend on the geometry of the structure, the magnitude of the displacements, and the frequency of resonance. For example:

Stroboscopy. This method is used when the displacements are large enough to be seen by the human eye and the frequency is higher than about twenty hertz.

Powder. In this method a thin layer of dry powder is sprinked on a structure whose geometry is a horizontal, planar surface. Under vibration, the grains of powder migrate to the nodal lines. It is necessary that the acceleration exceeds that of gravity. Sand is often used for this purpose.

Phase measurement. A nodal line separates two regions of opposite phase and this can be detected by comparing the phases of two transducers. One transducer is used as a fixed reference and the other is moved around the structure. At each point where the phase shifts by 180°, a nodal line has been crossed.

Amplitude minima. The usual procedure is to use an accelerometer attached to the top of a hand-held probe. The probe can then be used to search the structure surface for zones of minimum acceleration amplitude.

Holographic interferometry. This method can either be done in a time-averaged mode or in a real-time mode. The latter allows the complete pattern of nodal lines to be depicted at a given instant of time.

Of course, the shape and distribution of nodal lines is only a qualitative indication of the mode. A quantitative determination is usually done by using an array of piezoelectric accelerometers. Alternatively, one can use one fixed reference accelerometer and one moving accelerometer; this requires only two accelerometers, but more test time.

6.3.2 Spectral analysis

Spectral analysis is the process of representing a general function of time in terms of the amplitude and phase of a series of harmonic functions. In vibration, the general function is an analog signal which can be, for example, a force, an acceleration, or a ratio of a force over an acceleration. This spectral representation is a frequency-domain description of the time-dependent general function $x(t)$.

If $x(t)$ is the given time function, the Fourier transform establishes the relationship between $x(t)$ and its frequency-domain representation $X(f)$, i.e.

$$X(f) = \int_{-\infty}^{\infty} x(t)e^{-j2\pi ft}\, dt \tag{21}$$

In (21), f is the frequency in hertz and $X(f)$ is a complex function of f. A schematic of the relationship between $x(t)$ and $X(f)$ is presented in Figure 12.

In practical situations, $x(t)$ is given in the form of an analog signal and not as an analytical formula. Furthermore, it is known only in a finite time interval (t_1, t_2); therefore, instead of using (21), it is necessary to use the discrete form of the Fourier transform; namely:

$$X\left(\frac{n}{NT}\right) = T \sum_{k=0}^{N-1} x(kT)e^{-j2\pi n(k/N)} \tag{22}$$

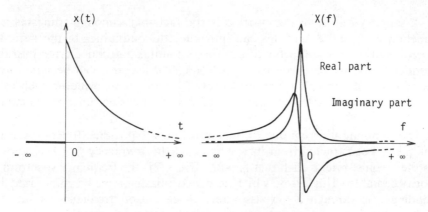

Figure 12   A function and its Fourier transform

where $n = 0, 1, \ldots, N-1$, and:

N, total number samples of x in the time domain, which also equals the total number of spectral lines of X in the frequency domain
T, the time interval between two successive samples of x
n, the numbers assigned to the spectral lines of X.

More information on the discrete Fourier transform (DFT) and on the fast Fourier transform (FFT) can be found for example in *The Fast Fourier Transform*, E. O. Brigham, Prentice-Hall, 1974. The relationship between a time-sampled function and its discrete frequency spectrum is shown schematically in Figure 13.

For a finite interval of time NT, it is first necessary to sample $x(t)$ in order to obtain N numerical values of $x(t)$ at the discrete times $t_1, t_1 + T, t_1 + 2T, \ldots, t_2$. Then the DFT can be used to obtain the corresponding discrete spectrum. This spectrum is the frequency-domain representation of a periodic function of period NT which is an approximation to $x(t)$ in the interval (t_1, t_2).

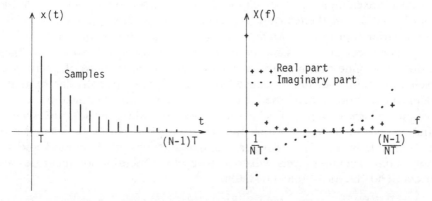

Figure 13   A sampled function and its discrete Fourier transform

A source of error in this process is the fact that sampling truncates the function at t_1 and t_2 and this can introduce discontinuities in the periodic representation of the function. Such discontinuities introduce error into the frequency spectrum. This error, which is called leakage, can be minimized by tapering the sampled amplitudes $x(nT)$ with a data window such as a triangular function (Bartlett window) or a shifted cosine function (Hanning window).

The sampling theorem due to Shannon states that a time function can be uniquely reconstructed from its spectrum, if the frequency $f_n = 1/T$, known as the Nyquist rate, is such that $f_n \geqslant 2f_c$. If $f_n < 2f_c$ the frequency spectrum is not acceptable. This error, which is called aliasing, can be minimized by sending $x(t)$ through a lowpass filter whose cut-off frequency is set $\leqslant f_c$ before the signal is sampled. These are called anti-aliasing filters. In practice, one usually selects a sampling rate so that $f_n = \alpha f_c$, where α is in the range of 3 to 4.

If the signal represents a displacement or a force at a point of a structure which is subjected to its actual operating conditions, real-time spectral analysis allows an immediate determination of the frequencies at which the structure has large response amplitudes. This then provides insight into either excitation frequencies which must be avoided, or frequencies which can be the resonant frequencies of the structure.

White noise is often used as a convenient form of broadband excitation. White noise is a random analog signal whose spectrum is continuous and constant over a range of frequencies; if the range of frequencies is broad, the white noise is broadband. The response of a linear structure to a white noise excitation is the superposition of the responses to all the frequencies in the excitation bandwidth. Spectral analysis of this response will give the same information as sinusoïdal sweeping but in this case all the frequencies are excited simultaneously and the data is processed in real-time. However, the response to a single sample of broadband excitation can be corrupted with noise; this noise can be minimized by averaging over many samples but this tends to offset the time advantage gained by real-time spectral analysis.

Impact excitation can be imposed by hammer blows and as a first approximation, mathematically represented by a Dirac function of force. The Fourier transform of a Dirac function gives a continuous and constant frequency spectrum, the same type of spectrum associated with broadband white noise. One can then utilize a shock excitation in place of white noise excitation. Shock excitation caused by impact with specially instrumented hammers is now widely used by specialists in the identification of the dynamic characteristics of structures. As for white noise, the spectral analysis of the response from impact excitation can be carried out in real-time. However, control over the signal bandwidth is limited and too high initial excitation levels will tend to excite structural nonlinearities and to overload the measurement system.

The ability to generate frequency spectrums rapidly has not only permitted real-time analysis for measurement purposes but it has facilitated the development of the field of machinery health monitoring. A single instrument can repeatedly scan a large number of transducers and compute a spectrum analysis of the signal for each one. These spectra can then be stored digitally and compared with their predecessors so that changes in the vibration signal can be identified. Such changes are often the precursors to failure or serious performance degradation.

7
Computer Programs

The computer programs presented here can be used on a microcomputer. They are useful for both beginners and experienced workers in mechanical vibration. For the former, these programs allow experience with the dynamic behavior of multi-degree of freedom structures without the labor of lengthy hand calculation. For the latter, they offer reference solutions, means of comparing various solution techniques, and the possibility of solving real problems on a desk-top computer.

The programs are written in BASIC and can be used on a computer which has at least 16K bytes of core memory. In addition, they use some specialized matrix commands and have the ability to present plots on a graphic terminal. They are easily adaptable to another language and their size can be greatly reduced by suppressing the comments and eliminating options, such as plotting.

When the algorithms have not been presented in the preceding chapters, they are briefly explained during the presentation of each program.

The contents of the chapter are as follows:

7.1   Language and Commentary

7.1.1   Language

The language used is standard BASIC. In addition, there are some matrix commands and plotting commands which are specific to the microcomputer used to develop the programs, and they will be briefly presented in this section.

Matrix commands

Addition, subtraction, and equality of matrices are done in the same way as for ordinary variables. Also, all the elements of a matrix can be set to the same value, including zero. In addition to the previous matrix commands, the following are used:

$$
\begin{array}{ll}
\text{CALL 'IDN', A} & A \text{ is the unit matrix} \\
\text{A = TRN (B)} & A \text{ is the transpose of } B \\
\text{A = B MPY C} & A \text{ is the product of } B \text{ and } C \\
\text{A = INV (A)} & A \text{ is the inverse of } A.
\end{array}
$$

Plotting commands

VIEWPORT R_1, R_2, S_1, S_2 defines the location and size of the rectangle in which the plot will be made as a proportion of the total surface available. The R and S axes are divided, respectively, into 130 and 100 graphic units. The abscissa R_1R_2 and the ordinate S_1S_2 define the graphic coordinates by the four corners of the rectangle; see Figure 1.

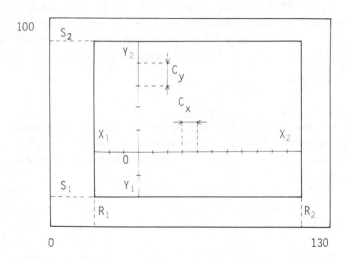

Figure 1   Notation used in plotting commands

WINDOW X_1, X_2, Y_1, Y_2 defines the scales of the plot in the rectangular area defined by VIEWPORT. X_1 and X_2 represent the extreme values of the X scale; Y_1 and Y_2 represent the extreme values of the Y scale.

AXIS C_x, C_y draws the axes O, X and O, Y. O is the origin of the axes defined by the command WINDOW. Each axis is divided into several parts by tic marks separated by distances equal to C_x and C_y in the units of the respective scale.

MOVE X, Y moves the pointer from its position to a point with the coordinates X, Y.

DRAW X, Y draws the plot from the position of the pointer to the point with coordinates X, Y.

7.1.2 Commentary

For the sake of simplicity and for the ease of comprehension, the use of core memory and computational time have not been optimized. For example, although many of the matrices used are banded, they are always treated as if they were full. This increases the size of memory used but simplifies the writing of matrix operations. Also, instead of using the best method to solve a particular set of linear equations, matrix inversion is always used.

Conversational mode

The program uses the conversational mode for data entry, for choosing different options, and for data modification if there are errors such as division by zero, matrix not defined, etc. It is not necessary to define the order of the matrix before execution. The order is determined automatically as the data is entered. If the user must choose between several options, the program writes a statement requesting numeric or alphanumeric input and the expected format of answers. When the elements of a symmetric matrix must be entered, only those of the upper triangle are requested.

Plotting

The plots are mostly modulus and phase of steady-state response to a harmonic excitation as a function of forcing frequency. There are also plots of the steady-state response of a system subjected to periodic excitation as a function of time.

The dimensions of the various plots are defined in the program and can only be changed by modifying the argument of the VIEWPORT command. In contrast, the scales must be defined in the course of the execution of the program. If the scale is logarithmic, it is defined by an integral number of decades and the value of the power of 10 at the origin. If the scale is linear, it is defined by its extreme values. After the definition of scales, the limits and sweep rate, in either time or frequency, are requested by the program. In some cases, the sweep rate is either linear or logarithmic. In order to facilitate the choice of plot scales, some information is printed; for example, resonant frequency or response maximum, unless their determination involves too much computation.

Calculation of phase

When one calculates the steady-state response of a system to a sinusoidal excitation of frequency Ω, the solution is given in terms of the amplitude X and the phase a, i.e.

$$x = X \sin (\Omega t - a)$$

All the calculations are made using complex notation:

$$x = I_m[(X_r + jX_i)e^{j\Omega t}]$$

Then

$$\sin a = -\frac{X_i}{X} \quad \text{and} \quad \cos a = \frac{X_r}{X}$$

with

$$X = (X_r^2 + X_i^2)^{1/2}$$

In the programs the phase is calculated from knowledge of X, $\cos a$ and $\sin a$ as follows.

If $X = 0$ then $a = 0$

If $X \neq 0$

         if $\cos a = 1$ $a = 0$

         if $\cos a = -1$ $a = 180$

         if $\cos a \neq \pm 1$ then:

            if $\sin a > 0$ $a = \dfrac{180}{\pi} \cos^{-1} a$

            if $\sin a < 0$ $a = \dfrac{180}{\pi} (2\pi - \cos^{-1} a)$

7.2   Presentation of Programs

Program 1: Steady-state response of a single degree-of-freedom system subjected to a sinusoidal force

The steady-state response is calculated using the relations developed in Chapter 1.

Options

- Viscous or structural damping
- Numerical listing of modulus and phase as a function of frequency
- Plot of modulus as a function of frequency
- Plot of phase as a function of frequency.

Program 2: Steady-state response of a single degree-of-freedom system subjected to a periodic force

The steady-state response is calculated for viscous damping. The force of excitation of period $T = 2\pi/\Omega$ must be developed in a Fourier series and the coefficients entered in the program. The differential equation of motion is

$$mx^{\infty} + cx^{\circ} + kx = \frac{a_0}{2} + \sum_{p=1}^{n} a_p \cos p\Omega t + \sum_{p=1}^{n} b_p \sin p\Omega t \tag{1}$$

If

$$R_p = \frac{k - mp^2\Omega^2}{(k - mp^2\Omega^2)^2 + c^2 p^2 \Omega^2} \tag{2}$$

and

$$I_p = \frac{-cp\Omega}{(k - mp^2\Omega^2)^2 + c^2 p^2 \Omega^2} \tag{3}$$

then

$$x = \frac{a_0}{2k} + \sum_{p=1}^{n} a_p (R_p \cos p\Omega t - I_p \sin p\Omega t)$$

$$+ \sum_{p=1}^{n} b_p (I_p \cos p\Omega t + R_p \sin p\Omega t) \tag{4}$$

Options

· Numerical listing of response as a function of time
· Plot of response as a function of time.

The plot is made only after the calculation and storage of all the response values. The extreme values of response are indicated before executing the plot in order to assist in the choice of the appropriate scales for the plot.

Program 3: Frequencies and modes of a spring–mass system by the transfer matrix method

The elements of the system must form a single chain and must be either springs or masses. The transfer matrices are calculated as presented in exercises 18 and 19 of Chapter 1. Application of the boundary conditions will require that a frequency dependent expression be equal to zero. The frequencies of free vibration are the roots of this expression; they are found by a frequency sweep.

Options

· Fixed–fixed system
· Fixed–free system
· Free–fixed system
· Free–free system
· Frequency sweep
· Calculation of the modes.

It is possible to redo a sweep or a mode calculation without re-entering the data which describes the system. Each boundary of the system, free or fixed, is numbered. The calculation of the modes begins with a frequency of free vibration and hence its accuracy is dependent on the accuracy of the frequency. Depending on whether the system is fixed or free at its first boundary, either the force or the displacement is set equal to unity. The components of the force or displacement of the other nodes are then determined with respect to this normalization.

Program 4: Steady-state response of a spring–mass–damper system subjected to a sinusoidal excitation by the transfer matrix method

The steady-state response is calculated for a system composed of a single chain of the following elements:
  · spring
  · mass
  · viscous damper
  · spring and viscous damper in parallel
  · spring having structural damping
  · sinusoidal force of excitation.

The transfer matrices are of the type presented in exercise 20 of Chapter 1. The modulus and phase of the response are calculated at the node chosen by the user. It is allowable to have several exciting forces but each force must be defined by its modulus and phase ϕ in the input data; for example, $F \cos \phi$ and $F \sin \phi$.

Options

  · Fixed–fixed system
  · Fixed–free system
  · Free–free system
  · Free–fixed system.

At the end of a sweep it is possible, without affecting the system data already entered, to:
  · change the boundary conditions
  · change the sweep
  · change the node at which the response is calculated.

Program 5: Frequencies and modes; modal masses and stiffnesses

This program calculates the frequencies and modes of a system of N degrees of freedom which has the equation of motion

$$Mx^{\infty} + Kx = 0 \qquad (5)$$

The mass matrix M and stiffness matrix K must be symmetric.

The iterative algorithm described in section 3.2 is used. The rigid-body modes are determined after modifying the matrix system as in section 5.2. In this procedure, the constant a must be chosen by the user. It is chosen to be zero when there are no rigid-body modes; if there are rigid-body modes, then the larger one chooses a, the slower the convergence of the iteration.

For each frequency calculation, the iterative procedure stops if the prescribed accuracy p is attained or if the number of iterations surpasses the prescribed maximum, n_{max}. If ω_j^r represents the jth frequency after the rth iteration, convergence is considered to be obtained if:

$$\frac{(\omega_j^{r+1})^2 - (\omega_j^r)^2}{(\omega_j^r)^2} < p \tag{6}$$

In the case where the accuracy p cannot be attained without exceding n_{max}, the program states this and allows the user to return to the beginning of the calculation without re-entering any system data, other than new values for p and n_{max}.

To be rigorous, the trial vector at the beginning of the iteration should have arbitrary elements. Instead the program uses an initial trial vector all of whose elements are unity. The final eigenvectors are normalized so that the largest element is unity.

In addition, it should be noted that the convergence is uncertain if the matrices M and K represent two or more uncoupled motions, e.g. bending and axial deformation of a straight member. On the other hand, the method and the computer program are limited to a reasonable number of degrees of freedom.

Program 6: Steady-state response of a damped system subjected to a sinusoidal excitation by the direct method

The system has N degrees of freedom and the mass matrix M and stiffness matrix K are symmetric. The damping need not be proportional. The steady-state response is calculated and the calculations are performed using the direct method described in section 3.3.

Options

- Viscous damping
- Structural damping
- Numerical listing of response modulus and phase as a function of frequency
- Plot of response modulus as a function of frequency
- Plot of response phase as a function of frequency.

At the end of a sweep or of a plot, it is possible to restart the calculations without re-entering any system data. The calculations are performed using complex notation and the forces of excitation have the form

$$(F_1 + jF_2)e^{j\Omega t} \tag{7}$$

The vectors F_1 and F_2 must be supplied as input data.

Program 7: Steady-state response of a damped system subjected to a sinusoidal excitation by the modal method

The response is calculated for n modes and uses the modal method described in section 3.3. The input data (modal masses, modal stiffnesses, modal damping, modes) can be calculated by using program 5. Complex notation is used and the generalized force vector must be calculated separately as shown in equation (64) of Chapter 3.

Options

- Viscous damping
- Structural damping
- Numerical listing of response modulus and phase as a function of frequency
- Plot of response modulus as a function of frequency
- Plot of response phase as a function of frequency.

At the end of a sweep or a plot, it is possible to restart the calculation without re-entering any system data.

Program 8: Response of a system to an arbitrary forcing function by the Wilson-θ method

The system has N degrees of freedom. The damping is viscous and the mass matrix and stiffness matrix are symmetric. The step-by-step Wilson θ method is used (see reference 1 in the Bibliography). The acceleration varies linearly from t to $t + \theta \Delta t$ with $\theta \geqslant 1$ and the motion at $t + \Delta t$ is calculated from a knowledge of the motion at t. For the choice $\theta = 1$, the method is identical to that presented in section 3.3. For solution stability, it is necessary to choose $\theta \geqslant 1.37$ and, in practice, it is suggested that one chooses $\theta = 1.40$.

Although at the initial instant $t = t_0$, the displacements, velocities, and accelerations are related, the program asks for all the initial data to be entered. At each instant of time, the program prints the values of displacement, velocity, and acceleration at each degree of freedom of the system. There is no plot of response. In order to ensure good results, it is suggested that the time increment be taken to be less than one-tenth of the smallest period of free vibration possessed by the system considered.

Program 9: Frequencies of a beam in bending, modal masses and stiffnesses, modes, deflection, slope, maximum stress

Equations (56) and (61) of Chapter 4 show that the frequencies ω_i and associated modes $\phi_i(x)$ are given by

$$\omega_i = \frac{X_i^2}{L^2} \sqrt{\frac{EI}{\rho S}} \tag{8}$$

and

$$\phi_i(x) = C_i \sin \beta_i x + D_i \cos \beta_i x + E_i \sinh \beta_i x + F_i \cosh \beta_i x \tag{9}$$

where

$$\beta_i = \sqrt[4]{\frac{\rho S \omega_i^2}{EI}} \tag{10}$$

The frequencies are calculated from (8) with the appropriate values of X_i selected from within the program depending on the boundary conditions. In the calculation of the modes, the constants C_i are arbitrarily set equal to unity and the constants D_i, E_i, F_i are obtained from the boundary conditions. Let

$$si = \sin \beta_i L \qquad sh = \sinh \beta_i L$$

$$co = \cos \beta_i L \qquad ch = \cosh \beta_i L$$

then the expressions for the constants are as follows:

System C–F: $\quad C_i = 1 \quad D_i = -\dfrac{si + sh}{co + ch} \quad E_i = -1 \quad F_i = -D_i$

System C–C: $\quad C_i = 1 \quad D_i = \dfrac{sh - si}{co - ch} \quad E_i = -1 \quad F_i = -D_i$
or C–S

System F–F: $\quad C_i = 1 \quad D_i = \dfrac{sh - si}{co - ch} \quad E_i = 1 \quad F_i = D_i$

System F–S: $\quad C_i = 1 \quad D_i = -\dfrac{si + sh}{co + ch} \quad E_i = 1 \quad F_i = D_i$

System S–S: $\quad C_i = 1 \quad D_i = 0 \quad E_i = 0 \quad F_i = 0$

With knowledge of the constants C_i, D_i, E_i, F_i for each ω_i the slope is determined by differentiation and the maximum stress in a cross section is

$$\sigma_{max} = E \frac{d}{2} \frac{d^2 \phi_i(x)}{dx^2} \tag{11}$$

where d is the distance of the furthest point from the neutral axis. The modal masses are given by

$$m_i = \int_0^L \rho S \phi_i^2(x) \, dx \tag{12}$$

and the modal stiffnesses are calculated from ω_i and m_i.

Options

- Frequencies
- Modes, modal masses and stiffnesses
- Automatic or nonautomatic determination of the number of points used in the mode.

Program 10: Frequencies of a system composed of:
· beams
· supported inertias
· suspended inertias
by the transfer matrix method

The various elements which compose the system are either beams, inertias supported on springs, or inertias suspended by springs from the beam, as shown in Figure 2.

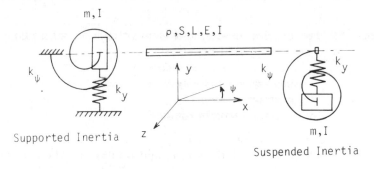

Figure 2   Transfer matrix elements used in program no. 10

The transfer matrix of a beam element in bending is given in exercise 10 of Chapter 4. The cross-section and properties must be constant along the length of the element. The length, however, can be arbitrary. The inertias must be masses m which move only in the y direction and rotatory inertias which rotate only about z. The deflectional stiffnesses k_y act in the y direction and the torsional stiffnesses k_ψ act around the z axis. The transfer matrices for the nonbeam elements can be calculated as in exercise 19 of Chapter 1, because the motions along y and around z are uncoupled. The frequencies are determined by requiring the determinant established by the boundary conditions to vanish. The frequencies are found by a frequency sweep, by observing the change in sign of the value of the determinant.

Options

- C–C system
- C–F system
- C–S system
- F–F system
- F–S system
- S–S system
- Frequency sweep
- Modification of boundary conditions.

Caution

- The transfer matrix of a beam is not defined for a zero frequency.
- One term of the transfer matrix of the suspended inertia contains $k_y - \Omega^2 m$ in the denominator (see the computer program). When the frequency Ω of the sweep is very close to $\sqrt{k_y/m}$, this term changes from $+\infty$ to $-\infty$. The sign of the value of the determinant can then change during the sweep while the value of the determinant does not vanish. In this case the corresponding value Ω is not a resonant frequency.

Program 11: Frequencies, modes, and element energies of a plane frame
composed of:
- **beams and bars elements**
- **mass–spring–mass systems**
- **arbitrary elements**
by the finite element method

All the elements in this program have two nodes with three degrees of freedom at each node: translations u, v, and rotation ψ. The coordinates of the two nodes i and j in the global frame of reference XYZ define the orientations of the local frame xyz. The x axis is oriented from i to j and the z axis is taken parallel to and in the same positive sense as Z; see Figure 3.

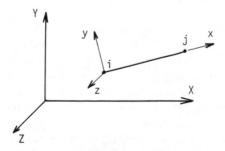

Figure 3 Global frame $(X\text{-}Y\text{-}Z)$ and local frame $(x\text{-}y\text{-}z)$ used in program no. 11.

Beam and bar element. These elements are straight members of constant cross section and characterized by E, ρ, S, I. The length is calculated by using the absolute coordinates of nodes i and j. The mass and stiffness matrices are determined in section 5.1. The mass matrix is the consistent mass matrix.

Mass–spring–mass system. The elements of these systems are defined by nodes i and j; one of these nodes need not be a node of the frame. Their energies are

$$T = \tfrac{1}{2}(m_{xi}u_i^{\circ 2} + m_{yi}v_i^{\circ 2} + m_{\psi i}\psi_i^{\circ 2} + m_{xj}u_j^{\circ 2} + m_{yj}v_j^{\circ 2} + m_{\psi j}\psi_j^{\circ 2}) \qquad (13)$$

$$U = \tfrac{1}{2}[k_x(u_i - u_j)^2 + k_y(v_i - v_j)^2 + k_\psi(\psi_i - \psi_j)^2] \qquad (14)$$

from which the mass and stiffness matrices are respectively written as

$$\begin{bmatrix} m_{xi} & & & & & 0 \\ & m_{yi} & & & & \\ & & m_{zi} & & & \\ & & & m_{xj} & & \\ & & & & m_{yj} & \\ \text{zero} & & & & & m_{zj} \end{bmatrix}; \qquad (15)$$

$$\begin{bmatrix} k_x & 0 & 0 & -k_x & 0 & 0 \\ 0 & k_y & 0 & 0 & -k_y & 0 \\ 0 & 0 & k_\psi & 0 & 0 & -k_\psi \\ -k_x & 0 & 0 & k_x & 0 & 0 \\ 0 & -k_y & 0 & 0 & k_y & 0 \\ 0 & 0 & -k_\psi & 0 & 0 & k_\psi \end{bmatrix} \qquad (16)$$

Arbitrary elements. These elements are characterized by full and symmetric mass and stiffness matrices. These matrices must be entered as data and must be defined in the local frame. Only one element of this type can be used.

Boundary conditions. It is necessary to enter the node numbers at which one or more of the displacements are set equal to zero. For each, one must enter the condition of zero displacement in the form: 0 for a free displacement and 1 for a zero displacement. The program eliminates the rows and columns corresponding to zero displacement to reduce the size of the matrices used in the eigenvalue algorithm. This algorithm is the same as that presented in program no. 5. The modes are expressed in the absolute frame of reference.

Element energies. After calculating the frequencies, modes, modal masses, and modal stiffnesses, the program calculates the percentage contribution of each element at the total kinetic and strain energy for each mode.

This program uses the most memory of any of the program given in this chapter. It requires three files which are used for: construction and assembly of the system matrices; calculation of the frequency, modes and element energies; and a file to safeguard the element matrices data.

Program 12: Gauss–Legendre integration

This program permits the calculation of the following integral:

$$\int_{x_1}^{x_2}\int_{y_1}^{y_2}\int_{z_1}^{z_2} f(x, y, z)\,dx\,dy\,dz \tag{17}$$

The method uses the Gauss–Legendre quadrature formulas given in Stroud, A. M., and Secrest, D., *Gaussian Quadrature Formulas*, Prentice-Hall (1966). It uses three sets of integration points ξ, η, ζ:

$$\int_{-1}^{+1}\int_{-1}^{+1}\int_{-1}^{+1} f(\xi, \eta, \zeta)\,d\xi\,d\eta\,d\zeta = \sum_{i=1}^{N}\sum_{j=1}^{N}\sum_{k=1}^{N} H_i H_j H_k f(\xi_i, \eta_i, \zeta_i) \tag{18}$$

The coordinates ξ_i, η_i, ζ_i are the so-called Gauss points and the quantities H_i, H_j, and H_k are weighting factors. It can be shown that a polynomial of degree $2N-1$ is exactly integrated by using N Gauss points. In the program, it is possible to use from 2 to 10 Gauss points and the function f can be in terms of one, two, or three variables. The change of units of integration is made automatically. In the course of execution, the program asks for the function f in the following manner:

- PROGRAM THE FUNCTION BEGINNING ON LINE 3000
- NUMBER THE LINES IN STEPS OF 10
- END THE NUMBERING WITH 'RETURN'
- TYPE ON THE KEYBOARD: RUN 5000

As an example, the function

$$f(x, y) = (x^2 - 4^2)^2 y^5$$

will be integrated. It is sufficient to type

$$3000 \qquad F = (X\uparrow 2 - 4\uparrow 2)\uparrow 2 * Y\uparrow 5$$

$$3010 \qquad RETURN$$

and execution will begin on the command RUN 5000.

This method can be used in approximate calculations such as the Rayleigh–Ritz method. It is also widely used in the evaluation of mass and stiffness matrices in the finite element method.

Options

- Number of variables in the function
- Number of Gauss points
- Modification of the function
- Change in the number of Gauss points
- Change in the limits of integration.

The last two options do not affect the function *f*.

Listings

```
100 REM                 PROGRAM NUMBER : 1
110 INIT
120 PAGE
130 PRINT  "*******************************************************"
140 PRINT  "*                                                     *"
150 PRINT  "*                STEADY STATE RESPONSE                *"
160 PRINT  "*        OF A SINGLE DEGREE-OF-FREEDOM SYSTEM         *"
170 PRINT  "*          SUBJECTED TO A SINUSOIDAL FORCE            *"
180 PRINT  "*                                                     *"
190 PRINT  "*         MASS=M                                      *"
200 PRINT  "*         STIFFNESS=K                                 *"
210 PRINT  "*         COEFFICIENT OF VISCOUS DAMPING=C            *"
220 PRINT  "*         STRUCTURAL DAMPING FACTOR=ETA               *"
230 PRINT  "*         FORCE AMPLITUDE=F                           *"
240 PRINT  "*                                                     *"
250 PRINT  "*     CALCULATION AND PLOTTING OF :X: AND a WITH      *"
260 PRINT  "*                x=:X:SIN(Wt-a)                       *"
270 PRINT  "*                                                     *"
280 PRINT  "*******************************************************"
290 DELETE Y,F3
300 PRINT
310 PRINT "MASS.......... M= ";
320 INPUT M
330 PRINT "STIFFNESS...... K= ";
340 INPUT K
350 PRINT "FORCE AMPLITUDE F= ";
360 INPUT F0
370 PRINT
380 PRINT "IF THE DAMPING IS VISCOUS        REP=VIS"
390 PRINT "          ''         IS STRUCTURAL   REP=STR"
400 PRINT "   REP= ";
410 INPUT B$
420 IF B$="VIS" THEN 450
430 IF B$="STR" THEN 500
440 GO TO 380
450 PRINT
460 PRINT "COEFFICIENT OF VISCOUS DAMPING  C= ";
470 INPUT C
480 C2=1
490 GO TO 540
500 PRINT
510 PRINT "STRUCTURAL DAMPING FACTOR  ETA= ";
520 INPUT E
530 C2=2
540 PRINT
550 PRINT "DO YOU WISH THE RESPONSE IN NUMERICAL FORM           REP=NUM"
560 PRINT "              ''    TO PLOT THE AMPLITUDE OF THE RESPONSE  REP=AMP"
```

```
570 PRINT "              ''   TO PLOT THE PHASE OF THE RESPONSE          REP=PHA"
580 PRINT "              ''   TO STOP...                                 REP=STO
590 PRINT
600 PRINT "  REP= ";
610 INPUT A$
620 IF A$="NUM" THEN 670
630 IF A$="AMP" THEN 690
640 IF A$="PHA" THEN 710
650 IF A$="STOP" THEN 2580
660 GO TO 540
670 C1=1
680 GO TO 750
690 C1=2
700 GO TO 750
710 C1=3
720 GO TO 750
730 C1=4
740 GO TO 1870
750 GO TO C1 OF 760,1310,2040,2580
760 PRINT
770 PRINT
780 PRINT "SWEEPING"
790 PRINT "----------------"
800 PRINT
810 F4=1/2/PI*SQR(K/M)
820 PRINT USING 830:"FREQUENCY OF THE UNDAMPED SYSTEM = ",F4," HERTZ"
830 IMAGE 37A,3E,6A
840 PRINT
850 PRINT "INITIAL FREQUENCY OF THE SWEEP   (HERTZ) = ";
860 INPUT F1
870 PRINT "FINAL                ''              ''     = ";
880 INPUT F2
890 PRINT "INCREMENT            ''              ''     = ";
900 INPUT F3
910 GO TO C1 OF 920,1510,2130,2580
920 PAGE
930 F=F1
940 W=2*PI*F
950 GO TO C2 OF 960,1020
960 D=(K-M*W*W)^2+(C*W)^2
970 O1=(K-M*W*W)/D^0.5
980 O1=O1-O1/1.0E+9
990 I1=+C*W/D^0.5
1000 M1=FO/SQR(D)
1010 GO TO 1070
1020 D=(K-M*W*W)^2+(E*K)^2
1030 O1=(K-M*W*W)/D^0.5
1040 O1=O1-O1/1.0E+9
1050 I1=+E*K/D^0.5
1060 M1=FO/SQR(D)
1070 IF M1<>0 THEN 1160
1080 T1=0
1090 GO TO 1200
1100 IF O1<>-1 THEN 1130
1110 T1=PI
1120 GO TO 1200
1130 IF O1<>1 THEN 1160
1140 T1=0
1150 GO TO 1200
1160 IF I1<0 THEN 1190
1170 T1=ACS(O1)
1180 GO TO 1200
1190 T1=2*PI-ACS(O1)
```

```
1200 T3=T1/PI*180
1210 GO TO C1 OF 1220,2000,2540,2580
1220 PRINT USING 1230:"FREQUENCY= ",F," |DISP|= ",M1," PHASE=",T3," DEG"
1230 IMAGE 11A,3E,9A,3E,8A,4D.2D,4A
1240 F=F+F3
1250 IF F>F2 THEN 1270
1260 GO TO 940
1270 PRINT
1280 GO TO C1 OF 1300,1290,1290,2580
1290 MOVE X1,Y1-10*T2
1300 GO TO 540
1310 PRINT
1320 PRINT
1330 PRINT "DEFINITION OF THE SCALE FOR THE PLOT OF MODULUS"
1340 PRINT "---------------------------------------------------"
1350 PRINT "MINIMUM VALUE OF THE FREQUENCY   (HERTZ) = ";
1360 INPUT X1
1370 PRINT "MAXIMUM              ''                  ''      = ";
1380 INPUT X2
1390 PRINT
1400 PRINT "HOW MANY DECADES DO YOU WISH ON THE ORDINATE : ";
1410 INPUT D2
1420 PRINT "                                  N"
1430 PRINT "ORIGIN OF THE ORDINATE   (10   MKSA)   N= ";
1440 INPUT P2
1450 IF P2=INT(P2) THEN 1500
1460 PRINT
1470 PRINT "CAUTION   N MUST BE AN INTEGER"
1480 PRINT "-------------------"
1490 GO TO 1420
1500 GO TO 760
1510 PAGE
1520 Y1=P2
1530 Y2=Y1+D2
1540 VIEWPORT 20,115,28,98
1550 T1=(X2-X1)/100
1560 T2=(Y2-Y1)/100
1570 T4=5*T1
1580 WINDOW X1-5*T1,X2+5*T1,Y1-T2,Y2+T2
1590 MOVE X1,Y1
1600 DRAW X2,Y1
1610 MOVE X1,Y1
1620 FOR I=1 TO 19
1630 MOVE X1+I*5*T1,Y1
1640 DRAW X1+I*5*T1,Y1+2*T2
1650 NEXT I
1660 MOVE X1+20*5*T1,Y1
1670 DRAW X1+20*5*T1,Y2
1680 MOVE X1-10*T1,Y1-5*T2
1690 PRINT USING 1700:X1
1700 IMAGE 2E
1710 MOVE X1+15*T1,Y1-5*T2
1720 PRINT USING 1730:"FREQUENCY INTERVAL= ",T4
1730 IMAGE 20A,2E
1740 MOVE X2-20*T1,Y1-5*T2
1750 PRINT USING 1760:X2,"   HERTZ"
1760 IMAGE 2E,7A
1770 GO TO C1 OF 1780,1780,2040,2580
1780 MOVE X1,Y1
1790 DRAW X1,Y2
1800 MOVE X1-8*T1,Y1
1810 PRINT "10"
1820 MOVE X1-6*T1,Y1+3*T2
```

```
1830 PRINT Y1
1840 MOVE X1,Y1
1850 FOR I=Y1 TO Y2-1
1860 FOR J=1 TO 8
1870 MOVE X1,I+LGT(J+1)
1880 DRAW X1+2*T1,I+LGT(J+1)
1890 NEXT J
1900 MOVE X1,I+1
1910 DRAW X2,I+1
1920 MOVE X1-8*T1,I+1
1930 PRINT "10"
1940 MOVE X1-6*T1,I+1+3*T2
1950 PRINT I+1
1960 NEXT I
1970 MOVE X1-8*T1,Y2-6*T2
1980 PRINT "AMP"
1990 GO TO 930
2000 IF F>F1 THEN 2020
2010 MOVE F1,LGT(M1)
2020 DRAW F,LGT(M1)
2030 GO TO 1240
2040 PRINT
2050 PRINT
2060 PRINT "DEFINITION OF THE SCALE FOR THE PLOT OF PHASE"
2070 PRINT "------------------------------------------------"
2080 PRINT "MINIMUM VALUE OF THE FREQUENCY   (HERTZ) = ";
2090 INPUT X1
2100 PRINT "MAXIMUM              ''              ''    = ";
2110 INPUT X2
2120 GO TO 760
2130 PAGE
2140 Y1=0
2150 Y2=180
2160 VIEWPORT 20,115,28,98
2170 T1=(X2-X1)/100
2180 T2=(Y2-Y1)/100
2190 T4=5*T1
2200 WINDOW X1-5*T1,X2+5*T1,Y1-T2,Y2+T2
2210 MOVE X1,Y1
2220 DRAW X2,Y1
2230 MOVE X1,Y1
2240 FOR I=1 TO 19
2250 MOVE X1+I*5*T1,Y1
2260 DRAW X1+I*5*T1,Y1+2*T2
2270 NEXT I
2280 MOVE X1+20*5*T1,Y1
2290 DRAW X1+20*5*T1,Y2
2300 MOVE X1-10*T1,Y1-5*T2
2310 PRINT USING 1700:X1
2320 IMAGE 2E
2330 MOVE X1+15*T1,Y1-5*T2
2340 PRINT USING 1730:"FREQUENCY INTERVAL= ",T4
2350 IMAGE 20A,2E
2360 MOVE X2-20*T1,Y1-5*T2
2370 PRINT USING 1760:X2,"  HERTZ"
2380 IMAGE 2E,7A
2390 MOVE X1,Y1
2400 DRAW X1,Y2
2410 Y3=0
2420 MOVE X1-10*T1,Y1
2430 PRINT Y3
2440 FOR I=1 TO 2
2450 Y3=Y3+90
```

```
2460 MOVE X1-10*T1,Y3
2470 PRINT Y3
2480 MOVE X1,Y3
2490 DRAW X2,Y3
2500 NEXT I
2510 MOVE X1-10*T1,Y2-6*T2
2520 PRINT "PHASE"
2530 GO TO 930
2540 IF F>F1 THEN 2560
2550 MOVE F1,T3
2560 DRAW F,T3
2570 GO TO 1240
2580 END


100 REM               PROGRAM NUMBER : 2
110 INIT
120 PAGE
130 PRINT "************************************************"
140 PRINT "*                                              *"
150 PRINT "*            STEADY STATE RESPONSE             *"
160 PRINT "*   OF A SINGLE DEGREE-OF-FREEDOM SYSTEM       *"
170 PRINT "*       SUBJECTED TO A PERIODIC FORCE          *"
180 PRINT "*                                              *"
190 PRINT "*         M.X'' + C.X' + K.X = F(t)            *"
200 PRINT "*                                              *"
210 PRINT "************************************************"
220 DELETE X,T4,A,B
230 PRINT
240 PRINT "MASS= ";
250 INPUT M
260 PRINT "STIFFNESS= ";
270 INPUT K
280 PRINT "COEFFICIENT OF VISCOUS DAMPING C= ";
290 INPUT C
300 PRINT
310 PRINT
320 PRINT "                      N"
330 PRINT " IF F(t)= A0/2 + SIGMA [A(P).COS(P.W.t) + B(P).SIN(P.W.t)]"
340 PRINT "                     P=1"
350 PRINT
360 PRINT "N= ";
370 INPUT N
380 IF N=0 THEN 400
390 DIM A(N),B(N)
400 PRINT
410 PRINT "FREQUENCY OF EXCITATION (HERTZ)= ";
420 INPUT F8
430 W=2*PI*F8
440 PRINT
450 PRINT "A0= ";
460 INPUT A0
470 PRINT
480 IF N=0 THEN 590
490 FOR I=1 TO N
500 PRINT "A(";I;")= ";
510 INPUT A(I)
520 NEXT I
530 PRINT
540 FOR I=1 TO N
550 PRINT "B(";I;")= ";
560 INPUT B(I)
570 NEXT I
```

```
580 PRINT
590 PRINT
600 PRINT "SWEEPING"
610 PRINT "-------------------"
620 TO=2*PI/W
630 PRINT USING 640:"THE PERIOD OF THE EXCITATION IS : ",TO," SEC"
640 IMAGE 34A,4E,4A
650 PRINT
660 PRINT "INITIAL TIME OF THE SWEEP  (SECOND) = ";
670 INPUT T1
680 PRINT "FINAL            ''              ''    = ";
690 INPUT T2
700 PRINT "INCREMENT        ''              ''    = ";
710 INPUT T3
720 PAGE
730 N1=(T2-T1)/T3+1
740 DIM X(N1),T4(N1)
750 J=0
760 Y1=1.0E+50
770 Y2=-1.0E+50
780 FOR T=T1 TO T2 STEP T3
790 J=J+1
800 X(J)=A0/K/2
810 T4(J)=T
820 IF N=0 THEN 930
830 FOR I=1 TO N
840 A1=I*W*T
850 CO=COS(A1)
860 SO=SIN(A1)
870 C1=K-M*I*I*W*W
880 C2=C*I*W
890 D=C1*C1+C2*C2
900 IF D=0 THEN 1610
910 X(J)=X(J)+A(I)*(C1*CO+C2*SO)/D+B(I)*(C1*SO-C2*CO)/D
920 NEXT I
930 X3=X(J)
940 Y1=Y1 MIN X3
950 Y2=Y2 MAX X3
960 GO TO 970
970 NEXT T
980 PRINT
990 PRINT "DO YO WISH A PLOT  YES OR NO : ";
1000 INPUT A$
1010 IF A$="YES" THEN 1190
1020 IF A$="NO" THEN 1040
1030 GO TO 980
1040 PAGE
1050 J=0
1060 FOR T=T1 TO T2 STEP T3
1070 J=J+1
1080 PRINT USING 1090:"t= ",T," SECOND"," X= ",X(J)
1090 IMAGE 3A,4E,7A,4A,4E
1100 NEXT T
1110 PRINT
1120 PRINT "DO YO WISH A PLOT  YES OR NO : ";
1130 INPUT A$
1140 IF A$="YES" THEN 1180
1150 IF A$="NO" THEN 1630
1160 GO TO 1110
1170 GO TO 1630
1180 PRINT
1190 PRINT
1200 PRINT "THE MAXIMUM CALCULATED VALUE OF X IS: ";Y2
```

```
1210 PRINT "   MINIMUM                    ''              : ";Y1
1220 PRINT
1230 PRINT "SCALE OF THE PLOTTING"
1240 PRINT "----------------------"
1250 PRINT "MAXIMUM VALUE OF X= ";
1260 INPUT Y2
1270 PRINT "MINIMUM   ''        = ";
1280 INPUT Y1
1290 PRINT "MINIMUM VALUE OF t= ";
1300 INPUT X1
1310 PRINT "MAXIMUM   ''        = ";
1320 INPUT X2
1330 PAGE
1340 VIEWPORT 20,115,10,100
1350 WINDOW X1,X2,Y1,Y2
1360 C1=(X2-X1)/20
1370 C2=(Y2-Y1)/20
1380 AXIS C1,C2
1390 MOVE X1-2*C1,Y1-C2
1400 PRINT USING 1410:X1
1410 IMAGE 2E
1420 MOVE X1+3*C1,Y1-C2
1430 PRINT USING 1440:"TIME INCREMENT=",C1
1440 IMAGE 15A,2E
1450 MOVE X2-5*C1,Y1-C2
1460 PRINT USING 1470:X2,"   SEC"
1470 IMAGE 2E,6A
1480 MOVE X1-4*C1,Y1+0.5*C2
1490 PRINT USING 1500:Y1
1500 IMAGE 2E
1510 MOVE X1-4*C1,Y2-C2
1520 PRINT USING 1530:Y2
1530 IMAGE 2E
1540 FOR I=1 TO J
1550 IF I=1 THEN 1580
1560 DRAW T4(I),X(I)
1570 GO TO 1590
1580 MOVE T4(1),X(1)
1590 NEXT I
1600 GO TO 1630
1610 PRINT "CAUTION  (K-M.P.W.P.W)^2 + (C.P.W)^2 =0"
1620 PRINT "----------"
1630 END


100 REM              PROGRAM NUMBER : 3
110 INIT
120 PAGE
130 PRINT "*********************************"
140 PRINT "*                               *"
150 PRINT "*     FREQUENCIES AND MODES     *"
160 PRINT "*    OF A SPRING-MASS SYSTEM     *"
170 PRINT "*              BY                *"
180 PRINT "*    THE TRANSFER MATRIX METHOD  *"
190 PRINT "*                               *"
200 PRINT "*********************************"
210 PRINT
220 PRINT "TYPE OF ELEMENTS"
230 PRINT "----------------"
240 PRINT "SPRING   REP=1"
250 PRINT "MASS     REP=2"
260 PRINT
270 PRINT "NUMBER OF ELEMENTS= ";
```

```
280 INPUT N
290 M9=0
300 K9=0
310 DELETE C,T,T1,A
320 DIM C(N),T(2,2),T1(2,2),A(2,2)
330 PRINT
340 FOR I=1 TO N
350 PRINT "TYPE OF THE ELEMENT ";I;" REP= ";
360 INPUT C(I)
370 IF C(I)=1 THEN 390
380 IF C(I)=2 THEN 410
390 K9=K9+1
400 GO TO 430
410 M9=M9+1
420 GO TO 430
430 NEXT I
440 PRINT
450 PRINT
460 PRINT "IF THE SYSTEM IS   C-C   REP=1"
470 PRINT "              ''        C-F   REP=2"
480 PRINT "              ''        F-C   REP=3"
490 PRINT "              ''        F-F   REP=4"
500 PRINT
510 PRINT
520 PRINT "  REP= ";
530 INPUT C8
540 PRINT
550 DELETE M1,K1,F1
560 IF M9=0 THEN 630
570 DIM M1(M9)
580 FOR I=1 TO M9
590 PRINT "MASS    NU:";I;" M= ";
600 INPUT M1(I)
610 NEXT I
620 PRINT
630 IF K9=0 THEN 690
640 DIM K1(K9)
650 FOR I=1 TO K9
660 PRINT "SPRING NU:";I;" K= ";
670 INPUT K1(I)
680 NEXT I
690 PRINT
700 PRINT
710 PRINT "DO YOU WISH A SWEEP                REP=SWE"
720 PRINT "              ''    TO CALCULATE THE MODE   REP=MOD"
730 PRINT "              ''    TO STOP...            REP=STOP"
740 PRINT
750 PRINT "  REP= ";
760 INPUT A$
770 IF A$="SWE" THEN 810
780 IF A$="MOD" THEN 830
790 IF A$="STOP" THEN 1770
800 GO TO 690
810 C7=1
820 GO TO 840
830 C7=2
840 GO TO C7 OF 850,960,1770
850 PRINT
860 PRINT "SWEEPING"
870 PRINT "----------------"
880 PRINT
890 PRINT "INITIAL FREQUENCY OF THE SWEEP  (HERTZ) = ";
900 INPUT F2
```

```
910 PRINT "FINAL                    ''                ''     = ";
920 INPUT F3
930 PRINT "INCREMENT                ''                ''     = ";
940 INPUT F4
950 GO TO 1290
960 PRINT
970 PRINT "CALCULATION OF THE MODE "
980 PRINT "-------------------------------------------"
990 PRINT
1000 PRINT "FREQUENCY= ";
1010 INPUT F
1020 M9=0
1030 K9=0
1040 W=2*PI*F
1050 W2=W^2
1060 PAGE
1070 PRINT USING 1080:"AT THE FREQUENCY OF ",F," HERTZ"
1080 IMAGE 20A,3E,6A
1090 PRINT
1100 DELETE V1,V2
1110 DIM V1(2),V2(2)
1120 V1=0
1130 C6=1
1140 GO TO C8 OF 1150,1150,1170,1170
1150 V1(1)=1
1160 GO TO 1180
1170 V1(2)=1
1180 PRINT USING 1190:"NODE NU: ",C6," FORCE= ",V1(1)," DISP= ",V1(2)
1190 IMAGE 9A,2D,8A,3E,7A,3E
1200 FOR I=1 TO N
1210 C6=C6+1
1220 C1=C(I)
1230 GO TO C1 OF 1720,1670
1240 V2=A MPY V1
1250 V1=V2
1260 PRINT USING 1190:"NODE NU: ",C6," FORCE= ",V1(1)," DISP= ",V1(2)
1270 NEXT I
1280 GO TO 690
1290 PAGE
1300 FOR F=F2 TO F3 STEP F4
1310 M9=0
1320 K9=0
1330 W=2*PI*F
1340 W2=W^2
1350 T=0
1360 FOR I=1 TO N
1370 C1=C(I)
1380 GO TO C1 OF 1720,1670
1390 REM
1400 IF I>1 THEN 1430
1410 T=A
1420 GO TO 1450
1430 T1=A MPY T
1440 T=T1
1450 NEXT I
1460 GO TO C8 OF 1470,1490,1510,1530
1470 X=T(2,1)
1480 GO TO 1600
1490 X=T(1,1)
1500 GO TO 1600
1510 X=T(2,2)
1520 GO TO 1600
1530 X=T(1,2)
```

```
1540 IF F<>0 THEN 1600
1550 PRINT
1560 PRINT USING 1570:"FREQUENCY= ",F," HERTZ    RIGID BODY MODE"
1570 IMAGE 11A,3E,25A
1580 PRINT
1590 GO TO 1620
1600 PRINT USING 1610:"FREQUENCY= ",F," HERTZ    DETERMINANT= ",X
1610 IMAGE 11A,3E,23A,3E
1620 NEXT F
1630 PRINT
1640 PRINT
1650 GO TO 690
1660 END
1670 REM MASS MATRIX
1680 M9=M9+1
1690 CALL "IDN",A
1700 A(1,2)=-M1(M9)*W2
1710 GO TO 1760
1720 REM STIFFNESS MATRIX
1730 K9=K9+1
1740 CALL "IDN",A
1750 A(2,1)=1/K1(K9)
1760 GO TO C7 OF 1390,1240,1770
1770 END


100 REM                 PROGRAM NUMBER : 4
110 INIT
120 PAGE
130 PRINT "*********************************************"
140 PRINT "*                                           *"
150 PRINT "*          STEADY STATE RESPONSE            *"
160 PRINT "*       OF A SPRING-MASS-DAMPER SYSTEM      *"
170 PRINT "*    SUBJECTED TO A SINUSOIDAL EXCITATION   *"
180 PRINT "*                   BY                      *"
190 PRINT "*        THE TRANSFER MATRIX METHOD         *"
200 PRINT "*                                           *"
210 PRINT "*********************************************"
220 PRINT
230 M9=0
240 K9=0
250 C9=0
260 V9=0
270 I9=0
280 F9=0
290 PRINT
300 PRINT "TYPE OF ELEMENTS"
310 PRINT "--------------------------"
320 PRINT "REP=1   SPRING"
330 PRINT "REP=2   MASS"
340 PRINT "REP=3   VISCOUS DAMPER"
350 PRINT "REP=4   SPRING + VISCOUS DAMPER IN PARALLEL"
360 PRINT "REP=5   SPRING + STRUCTURAL DAMPER IN PARALLEL"
370 PRINT "REP=6   SINUSOIDAL EXCITING FORCE"
380 PRINT
390 PRINT "NUMBER OF ELEMENTS=";
400 INPUT N
410 DELETE C,T,T1,T2,A
420 DIM C(N),T(5,5),T1(5,5),T2(5,5),A(5,5)
430 PRINT
440 FOR I=1 TO N
450 PRINT "TYPE OF ELEMENT ";I;" REP= ";
460 INPUT C(I)
```

```
470 IF C(I)=1 THEN 540
480 IF C(I)=2 THEN 560
490 IF C(I)=3 THEN 580
500 IF C(I)=4 THEN 600
510 IF C(I)=5 THEN 620
520 IF C(I)=6 THEN 640
530 GO TO 450
540 K9=K9+1
550 GO TO 650
560 M9=M9+1
570 GO TO 650
580 C9=C9+1
590 GO TO 650
600 V9=V9+1
610 GO TO 650
620 I9=I9+1
630 GO TO 650
640 F9=F9+1
650 NEXT I
660 IF F9<>0 THEN 710
670 PRINT
680 PRINT "CAUTION   THERE IS NO EXCITING FORCE"
690 PRINT "————————————"
700 GO TO 220
710 PRINT
720 DELETE K1,M1,C1,F1,F2,K2,C2,K3,E3
730 IF M9=0 THEN 800
740 DIM M1(M9)
750 FOR I=1 TO M9
760 PRINT "MASS    NU:";I;" M= ";
770 INPUT M1(I)
780 NEXT I
790 PRINT
800 IF K9=0 THEN 870
810 DIM K1(K9)
820 FOR I=1 TO K9
830 PRINT "SPRING NU:";I;" K= ";
840 INPUT K1(I)
850 NEXT I
860 PRINT
870 IF C9=0 THEN 940
880 DIM C1(C9)
890 FOR I=1 TO C9
900 PRINT "DAMPER NU:";I;" C= ";
910 INPUT C1(I)
920 NEXT I
930 PRINT
940 IF V9=0 THEN 1030
950 DIM K2(V9),C2(V9)
960 FOR I=1 TO V9
970 PRINT "SPRING + VISCOUS DAMPER IN PARALLEL. NU:";I;" K= ";
980 INPUT K2(I)
990 PRINT "SPRING + VISCOUS DAMPER IN PARALLEL. NU:";I;" C= ";
1000 INPUT C2(I)
1010 NEXT I
1020 PRINT
1030 IF I9=0 THEN 1120
1040 DIM K3(I9),E3(I9)
1050 FOR I=1 TO I9
1060 PRI "SPRING + STRUCTURAL DAMPER IN PARALLEL. K(1+jE) NU:";I;" K= ";
1070 INPUT K3(I)
1080 PRI "SPRING + STRUCTURAL DAMPER IN PARALLEL. K(1+jE) NU:";I;" E= ";
1090 INPUT E3(I)
```

```
1100 NEXT I
1110 PRINT
1120 DIM F1(F9),F2(F9)
1130 FOR I=1 TO F9
1140 PRINT "EXCITING FORCE NU:";I;" F.COS(FI)= ";
1150 INPUT F1(I)
1160 PRINT "EXCITING FORCE NU:";I;" F.SIN(FI)= ";
1170 INPUT F2(I)
1180 NEXT I
1190 PRINT
1200 PRINT
1210 PRINT "AT WHICH NODE DO YOU WISH THE RESPONSE : ";
1220 INPUT N1
1230 IF N1<1 OR N1>N+1 THEN 1200
1240 PRINT
1250 PRINT
1260 PRINT
1270 PRINT "IF THE SYSTEM IS C-C   REP=1"
1280 PRINT "             ''       C-F   REP=2"
1290 PRINT "             ''       F-F   REP=3"
1300 PRINT "             ''       F-C   REP=4"
1310 PRINT
1320 PRINT "  REP= ";
1330 INPUT C8
1340 PRINT
1350 PRINT "SWEEPING"
1360 PRINT "————————"
1370 PRINT "INITIAL FREQUENCY OF THE SWEEP  (HERTZ) = ";
1380 INPUT F3
1390 PRINT "FINAL              ''                ''    = ";
1400 INPUT F4
1410 PRINT "INCREMENT          ''                ''    = ";
1420 INPUT F5
1430 PAGE
1440 FOR F=F3 TO F4 STEP F5
1450 M9=0
1460 K9=0
1470 C9=0
1480 V9=0
1490 I9=0
1500 F9=0
1510 C6=1
1520 IF C6=N1 THEN 1540
1530 GO TO 1550
1540 CALL "IDN",T2
1550 W=2*PI*F
1560 W2=W^2
1570 T=0
1580 FOR I=1 TO N
1590 CALL "IDN",A
1600 C3=C(I)
1610 GO TO C3 OF 2550,2500,2600,2650,2720,2790
1620 REM
1630 IF I>1 THEN 1660
1640 T=A
1650 GO TO 1680
1660 T1=A MPY T
1670 T=T1
1680 C6=C6+1
1690 IF C6=N1 THEN 1710
1700 GO TO 1720
1710 T2=T
1720 NEXT I
```

```
1730 DELETE B,V1,V2,E1,E2
1740 DIM B(2,2),V1(5),V2(5),E1(2),E2(2)
1750 GO TO C8 OF 1760,1800,1840,1880
1760 REM C--C
1770 I1=2
1780 J1=1
1790 GO TO 1910
1800 REM C--F
1810 I1=1
1820 J1=1
1830 GO TO 1910
1840 REM F--F
1850 I1=1
1860 J1=2
1870 GO TO 1910
1880 REM F--C
1890 I1=2
1900 J1=2
1910 B(1,1)=T(I1,J1)
1920 B(1,2)=T(I1,J1+2)
1930 B(2,1)=T(I1+2,J1)
1940 B(2,2)=T(I1+2,J1+2)
1950 E1(1)=-T(I1,5)
1960 E1(2)=-T(I1+2,5)
1970 D=B(1,1)*B(2,2)-B(1,2)*B(2,1)
1980 IF D<>0 THEN 2030
1990 PRINT
2000 PRINT "CAUTION  THE SYSTEM IS INDETERMINATE"
2010 PRINT "------------------------"
2020 GO TO 1240
2030 E2(1)=(B(2,2)*E1(1)-B(1,2)*E1(2))/D
2040 E2(2)=(B(1,1)*E1(2)-B(2,1)*E1(1))/D
2050 V1=0
2060 V1(5)=1
2070 IF J1=1 THEN 2110
2080 V1(2)=E2(1)
2090 V1(4)=E2(2)
2100 GO TO 2130
2110 V1(1)=E2(1)
2120 V1(3)=E2(2)
2130 V2=T2 MPY V1
2140 M2=SQR(V2(2)^2+V2(4)^2)
2150 IF M2<>0 THEN 2180
2160 T3=0
2170 GO TO 2300
2180 C4=V2(2)/M2
2190 S4=-V2(4)/M2
2200 IF C4<>-1 THEN 2230
2210 T3=PI
2220 GO TO 2300
2230 IF C4<>1 THEN 2260
2240 T3=0
2250 GO TO 2300
2260 IF S4<0 THEN 2290
2270 T3=ACS(C4)
2280 GO TO 2300
2290 T3=2*PI-ACS(C4)
2300 T3=T3/PI*180
2310 PRINT USING 2320:"FREQ= ",F," |DISP|= ",M2," PHASE= ",T3," DEG"
2320 IMAGE 6A,3E,9A,3E,8A,4D.2D,4A
2330 NEXT F
2340 PRINT
2350 PRINT
```

```
2360 PRINT "DO YOU WISH TO CHANGE THE BOUNDARY CONDITIONS  REP=BOU"
2370 PRINT "               ''         THE SWEEPING           REP=SWE"
2380 PRINT "               ''         THE RESPONSE NODE       REP=NOD"
2390 PRINT "               ''         TO STOP...             REP=STOP"
2400 PRINT
2410 PRINT "  REP= ";
2420 INPUT A$
2430 IF A$="BOU" THEN 1250
2440 IF A$="SWE" THEN 1340
2450 IF A$="NOD" THEN 1200
2460 IF A$="STOP" THEN 2840
2470 GO TO 2340
2480 PRINT
2490 GO TO 960
2500 REM MASS MATRIX
2510 M9=M9+1
2520 A(1,2)=-M1(M9)*W2
2530 A(3,4)=A(1,2)
2540 GO TO 1620
2550 REM STIFFNESS MATRIX
2560 K9=K9+1
2570 A(2,1)=1/K1(K9)
2580 A(4,3)=A(2,1)
2590 GO TO 1620
2600 REM VISCOUS DAMPING MATRIX
2610 C9=C9+1
2620 A(4,1)=-1/C1(C9)/W
2630 A(2,3)=-A(4,1)
2640 GO TO 1620
2650 REM MATRIX OF SPRING + VISCOUS DAMPER IN PARALLEL
2660 V9=V9+1
2670 A(2,1)=K2(V9)/(K2(V9)^2+C2(V9)^2*W2)
2680 A(4,3)=A(2,1)
2690 A(4,1)=-C2(V9)*W/(K2(V9)^2+C2(V9)^2*W2)
2700 A(2,3)=-A(4,1)
2710 GO TO 1620
2720 REM MATRIX OF SPRING + STRUCTURAL DAMPER IN PARALLEL K(1+jE)
2730 I9=I9+1
2740 A(2,1)=1/(K3(I9)*(1+E3(I9)^2))
2750 A(4,3)=A(2,1)
2760 A(4,1)=-E3(I9)/(K3(I9)*(1+E3(I9)^2))
2770 A(2,3)=-A(4,1)
2780 GO TO 1620
2790 REM MATRIX OF FORCE
2800 F9=F9+1
2810 A(1,5)=-F1(F9)
2820 A(3,5)=-F2(F9)
2830 GO TO 1620
2840 END


100 REM              PROGRAM NUMBER :5
110 INIT
120 PAGE
130 PRINT "***********************************"
140 PRINT "*                                 *"
150 PRINT "*       FREQUENCIES AND MODES      *"
160 PRINT "*    MODAL MASSES AND STIFFNESSES  *"
170 PRINT "*                                 *"
180 PRINT "*     [M].(X)'' +[K].(X) ={0}      *"
190 PRINT "*                                 *"
200 PRINT "***********************************"
210 PRINT
```

```
220 PRINT "ORDER OF THE SYSTEM = ";
230 INPUT N
240 DELETE M,K
250 DIM M(N,N),K(N,N)
260 PRINT
270 FOR I=1 TO N
280 FOR J=I TO N
290 PRINT "M(";I;",";J;")= ";
300 INPUT M(I,J)
310 NEXT J
320 NEXT I
330 PRINT
340 FOR I=1 TO N
350 FOR J=I TO N
360 PRINT "K(";I;",";J;")= ";
370 INPUT K(I,J)
380 NEXT J
390 NEXT I
400 FOR I=1 TO N
410 FOR J=I TO N
420 M(J,I)=M(I,J)
430 K(J,I)=K(I,J)
440 NEXT J
450 NEXT I
460 PRINT
470 PRINT "NUMBER OF MODES REQUIRED        = ";
480 INPUT N1
490 K8=1
500 DELETE K1,D
510 DIM K1(N,N),D(N,N)
520 PRINT
530 PRINT "ACCURACY   ON THE EIGENVECTORS   = ";
540 INPUT P
550 PRINT
560 PRINT "MAXIMUM    NUMBER OF ITERATIONS  = ";
570 INPUT N3
580 PRINT
590 PRINT "ARE THERE RIGID BODY MODES  YES OR NO : ";
600 INPUT A$
610 IF A$="YES" THEN 640
620 IF A$="NO" THEN 670
630 GO TO 580
640 PRINT "   IF   K + ALPHA*M         ALPHA= ";
650 INPUT A
660 GO TO 680
670 A=0
680 PAGE
690 PRINT
700 FOR I=1 TO N
710 FOR J=1 TO N
720 K1(I,J)=K(I,J)+A*M(I,J)
730 NEXT J
740 NEXT I
750 K1=INV(K1)
760 D=K1 MPY M
770 DELETE K1,V,S1,L
780 DIM V(N,N1),S1(N,N),L(N1)
790 V=0
800 S1=D
810 DELETE X,F
820 DIM X(N),F(N)
830 W3=0
840 X=1
```

```
850 K9=0
860 K9=K9+1
870 F=S1 MPY X
880 F2=0
890 FOR I=1 TO N
900 IF ABS(F(I))-F2<0 THEN 930
910 F2=ABS(F(I))
920 F3=F(I)
930 NEXT I
940 W2=1/F3
950 L(K8)=W2-A
960 X=F/F3
970 IF K9-N3<=0 THEN 1100
980 PRINT
990 PRINT "THE SYSTEM DOES NOT CONVERGE AFTER : ";K9;"  ITERATIONS"
1000 PRINT
1010 PRINT "DO YOU WISH TO CHANGE THE NUMBER OF EIGENVECTORS"
1020 PRINT "                          THE ACCURACY ON THE EIGENVECTORS"
1030 PRINT "             ''           THE MAXIMUM NUMBER OF ITERATIONS"
1040 PRINT "             ''           ALPHA"
1050 PRINT "  YES OR NO : ";
1060 INPUT A$
1070 IF A$="YES" THEN 460
1080 IF A$="NO" THEN 1990
1090 GO TO 980
1100 IF ABS((W2-W3)/W2)<P THEN 1130
1110 W3=W2
1120 GO TO 860
1130 PRINT "EIGENVECTOR NU: ";K8;" CONVERGENCE AFTER : ";K9;
1140 PRINT " ITERATIONS"
1150 FOR I=1 TO N
1160 V(I,K8)=X(I)
1170 NEXT I
1180 IF K8=>N1 THEN 1530
1190 DELETE S1,X,F,T,T1
1200 N2=N-K8
1210 DIM T(N1,N),T1(N1,N)
1220 T=TRN(V)
1230 T1=T MPY M
1240 DELETE T,R,R1,R2
1250 DIM R(K8,K8),R1(K8,N2),R2(K8,N2)
1260 FOR I=1 TO K8
1270 FOR J=1 TO N
1280 IF J>K8 THEN 1310
1290 R(I,J)=T1(I,J)
1300 GO TO 1330
1310 J1=J-K8
1320 R1(I,J1)=T1(I,J)
1330 NEXT J
1340 NEXT I
1350 R=INV(R)
1360 R2=R MPY R1
1370 DELETE R,R1,S
1380 DIM S(N,N)
1390 CALL "IDN",S
1400 FOR I=1 TO K8
1410 S(I,I)=0
1420 FOR J=K8+1 TO N
1430 J1=J-K8
1440 S(I,J)=-R2(I,J1)
1450 NEXT J
1460 NEXT I
1470 DELETE R2
```

```
1480 DIM S1(N,N)
1490 S1=D MPY S
1500 DELETE S
1510 K8=K8+1
1520 GO TO 810
1530 PRINT
1540 PRINT "EIGENVALUES"
1550 PRINT
1560 FOR I=1 TO N1
1570 PRINT "L(";I;")= ";L(I)
1580 NEXT I
1590 PRINT
1600 PRINT
1610 PRINT "FREQUENCIES [RAD/SEC]"
1620 PRINT
1630 FOR I=1 TO N1
1640 PRINT "W(";I;")= ";ABS(L(I))^0.5
1650 NEXT I
1660 PRINT
1670 PRINT
1680 PRINT "FREQUENCIES [HERTZ]"
1690 PRINT
1700 FOR I=1 TO N1
1710 PRINT "F(";I;")= ";ABS(L(I))^0.5/2/PI
1720 NEXT I
1730 PRINT
1740 PRINT
1750 PRINT "MODES"
1760 PRINT
1770 FOR J=1 TO N1
1780 PRINT
1790 PRINT "V. P. NO: ";J
1800 FOR I=1 TO N
1810 PRINT "                ";V(I,J)
1820 NEXT I
1830 NEXT J
1840 DELETE S1,T1,R
1850 DIM S1(N,N1),T1(N1,N),R(N1,N1)
1860 S1=M MPY V
1870 T1=TRN(V)
1880 R=T1 MPY S1
1890 PRINT
1900 PRINT "MODAL MASS MATRIX"
1910 PRINT R
1920 PRINT
1930 S1=K MPY V
1940 R=T1 MPY S1
1950 PRINT "MODAL STIFFNESS MATRIX"
1960 PRINT R
1970 PRINT
1980 GO TO 1000
1990 END
```

```
100 REM               PROGRAM NUMBER : 6
110 INIT
120 PAGE
130 PRINT "****************************************************************"
140 PRINT "*                                                              *"
150 PRINT "*                 STEADY STATE RESPONSE                        *"
160 PRINT "*    OF A SYSTEM SUBJECTED TO A SINUSOIDAL EXCITATION          *"
170 PRINT "*                        BY                                    *"
180 PRINT "*                 THE DIRECT METHOD                            *"
190 PRINT "*                                                              *"
200 PRINT "*           (x)=IMAGINARY PART OF (z)                          *"
210 PRINT "*                                                              *"
220 PRINT "*                              jWt                             *"
230 PRINT "*                        z=Ze                                  *"
240 PRINT "*                                                              *"
250 PRINT "*      (-W*W[M] + jW[C] + [K])(z)                    jWt       *"
260 PRINT "*                      OR                 =((F1)+j(F2))e        *"
270 PRINT "*      (-W*W[M]   + [K1]+j[K2])(z)                             *"
280 PRINT "*                                                              *"
290 PRINT "*          [M],[C],[K],[K1],[K2] ARE SYMETRIC                  *"
300 PRINT "*                                                              *"
310 PRINT "*       THE PHASE a IS DEFINED BY x=!X! SIN(Wt-a)              *"
320 PRINT "*                                                              *"
330 PRINT "****************************************************************"
340 PRINT
350 PRINT
360 PRINT "NUMBER OF DEGREES OF FREEDOM = ";
370 INPUT N
380 DELETE M,K1,K2,A,X,B
390 N2=2*N
400 DIM M(N,N),K1(N,N),K2(N,N),A(N2,N2),X(N2),B(N2)
410 PRINT
420 PRINT
430 PRINT "MASS MATRIX [M]"
440 PRINT
450 FOR I=1 TO N
460 FOR J=I TO N
470 PRINT "M(";I;",";J;")= ";
480 INPUT M(I,J)
490 NEXT J
500 NEXT I
510 PRINT
520 PRINT "IF THE DAMPING IS VISCOUS      REP=VIS"
530 PRINT "              ''      STRUCTURAL  REP=STR"
540 PRINT "  REP= ";
550 INPUT C$
560 IF C$="VIS" THEN 610
570 IF C$="STR" THEN 590
580 GO TO 510
590 C9=2
600 GO TO 620
610 C9=1
620 PRINT
630 PRINT
640 GO TO C9 OF 650,670
650 PRINT "STIFFNESS MATRIX [K]"
660 GO TO 680
670 PRINT "STIFFNESS MATRIX [K1]"
680 PRINT
690 PRINT
700 FOR I=1 TO N
710 FOR J=I TO N
720 GO TO C9 OF 730,750
```

```
730 PRINT "K(";I;",";J;")= ";
740 GO TO 760
750 PRINT "K1(";I;",";J;")= ";
760 INPUT K1(I,J)
770 NEXT J
780 NEXT I
790 PRINT
800 PRINT
810 GO TO C9 OF 820,840
820 PRINT "DAMPING MATRIX [C]"
830 GO TO 850
840 PRINT "DAMPING MATRIX [K2]"
850 PRINT
860 PRINT
870 FOR I=1 TO N
880 FOR J=I TO N
890 GO TO C9 OF 900,920
900 PRINT "C(";I;",";J;")= ";
910 GO TO 930
920 PRINT "K2(";I;",";J;")= ";
930 INPUT K2(I,J)
940 NEXT J
950 NEXT I
960 FOR I=1 TO N
970 FOR J=I TO N
980 K1(J,I)=K1(I,J)
990 K2(J,I)=K2(I,J)
1000 M(J,I)=M(I,J)
1010 NEXT J
1020 NEXT I
1030 PRINT
1040 PRINT
1050 PRINT "EXCITATION FORCES   {F1} AND {F2}"
1060 PRINT
1070 PRINT
1080 FOR I=1 TO N
1090 PRINT "F1(";I;")= ";
1100 INPUT B(I)
1110 NEXT I
1120 PRINT
1130 FOR I=N+1 TO N2
1140 J=I-N
1150 PRINT "F2(";J;")= ";
1160 INPUT B(I)
1170 NEXT I
1180 PRINT
1190 PRINT
1200 PRINT "NUMBER OF THE DEGREE OF FREEDOM TO BE CALCULATED NU= ";
1210 INPUT N3
1220 IF N3<1 OR N3>N THEN 1190
1230 PRINT
1240 PRINT
1250 PRINT "DO YOU WISH THE RESPONSE IN NUMERICAL FORM           REP=NUM"
1260 PRINT "            ''   TO PLOT THE AMPLITUDE OF THE RESPONSE  REP=AMP"
1270 PRINT "            ''   TO PLOT THE PHASE OF THE RESPONSE      REP=PHA"
1280 PRINT "            ''   TO STOP...                            REP=STOP"
1290 PRINT
1300 PRINT "  REP= ";
1310 INPUT A$
1320 IF A$="NUM" THEN 1370
1330 IF A$="AMP" THEN 1390
1340 IF A$="PHA" THEN 1410
1350 IF A$="STOP" THEN 3850
```

```
1360 GO TO 1230
1370 C1=1
1380 GO TO 1450
1390 C1=2
1400 GO TO 1450
1410 C1=3
1420 GO TO 1450
1430 C1=4
1440 GO TO 1230
1450 GO TO C1 OF 1460,2280,3130,3850
1460 PRINT
1470 PRINT
1480 PRINT "SWEEPING"
1490 PRINT "--------------------"
1500 PRINT
1510 PRINT "INITIAL FREQUENCY OF THE SWEEP  (HERTZ) = ";
1520 INPUT F1
1530 PRINT "FINAL           ''              ''      = ";
1540 INPUT F2
1550 PRINT
1560 PRINT "DO YOU WISH A LINEAR        INCREMENT   REP=LIN"
1570 PRINT "          ''        LOGARITHMIC    ''      REP=LOG"
1580 PRINT "  REP= ";
1590 INPUT B$
1600 PRINT
1610 IF B$="LIN" THEN 1640
1620 IF B$="LOG" THEN 1670
1630 GO TO 1600
1640 PRINT "INCREMENT              ''             ''      = ";
1650 INPUT F3
1660 GO TO 1770
1670 IF F1<>0 THEN 1730
1680 PRINT
1690 PRINT "CAUTION   THE INITIAL FREQUENCY WAS ZERO"
1700 PRINT "--------------   ITS LOG IS NOT DEFINED"
1710 PRINT "DEFINED"
1720 GO TO 1470
1730 PRINT "NUMBER OF EQUAL DIVISIONS ALONG"
1740 PRINT "THE LOGARITHMIC SCALE BETWEEN F1 AND F2 = ";
1750 INPUT F3
1760 F3=10^((LGT(F2)-LGT(F1))/F3)
1770 GO TO C1 OF 1780,2530,3310,3850
1780 PAGE
1790 F=F1
1800 W=2*PI*F
1810 FOR I=1 TO N
1820 I1=I+N
1830 FOR J=1 TO N
1840 J1=J+N
1850 A(I,J)=K1(I,J)-W^2*M(I,J)
1860 A(I1,J1)=A(I,J)
1870 GO TO C9 OF 1880,1900
1880 A(I1,J)=K2(I,J)*W
1890 GO TO 1910
1900 A(I1,J)=K2(I,J)
1910 A(I,J1)=-A(I1,J)
1920 NEXT J
1930 NEXT I
1940 A=INV(A)
1950 X=A MPY B
1960 M1=SQR(X(N3)^2+X(N3+N)^2)
1970 IF M1<>0 THEN 2000
1980 T1=0
```

```
1990 GO TO 2130
2000 O1=X(N3)/M1
2010 O1=O1-O1/1.OE+9
2020 I1=-X(N3+N)/M1
2030 IF O1<>-1 THEN 2060
2040 T1=PI
2050 GO TO 2130
2060 IF O1<>1 THEN 2090
2070 T1=0
2080 GO TO 2130
2090 IF I1<0 THEN 2120
2100 T1=ACS(O1)
2110 GO TO 2130
2120 T1=2*PI-ACS(O1)
2130 T1=T1/PI*180
2140 GO TO C1 OF 2150,3090,3810,3850
2150 PRINT USING 2160: "FREQUENCY= ",F," |DISP|= ",M1," PHASE=",T1," DEG"
2160 IMAGE 11A,3E,9A,3E,8A,4D.2D,4A
2170 IF B$="LIN" THEN 2200
2180 F=F*F3
2190 GO TO 2210
2200 F=F+F3
2210 IF F>F2 THEN 2240
2220 GO TO 1800
2230 NEXT F
2240 PRINT
2250 GO TO C1 OF 1230,2260,2260,3850
2260 MOVE X1,Y1-10*T2
2270 GO TO 1230
2280 REM AXIS LOG-LOG
2290 PRINT
2300 PRINT
2310 PRINT "DEFINITION OF THE SCALE OF THE LOG-LOG AXIS"
2320 PRINT "................................................................................"
2330 PRINT
2340 PRINT "HOW MANY FREQUENCY DECADES DO YOU WISH  : ";
2350 INPUT D1
2360 PRINT "     ''     RESPONSE              ''        : ";
2370 INPUT D2
2380 PRINT
2390 PRINT "                                        N"
2400 PRINT "ORIGIN OF THE ABSCISSA  (10   HERTZ)    N= ";
2410 INPUT P1
2420 IF INT(P1)<>P1 THEN 2480
2430 PRINT
2440 PRINT "                                        N"
2450 PRINT "        ''       ORDINATE   (10   MKSA )   N= ";
2460 INPUT P2
2470 IF INT(P2)=P2 THEN 2520
2480 PRINT
2490 PRINT "CAUTION  N MUST BE AN INTEGER"
2500 PRINT "................"
2510 GO TO 2380
2520 GO TO 1460
2530 PAGE
2540 IF F1<>0 THEN 2580
2550 PRINT "CAUTION   THE INITIAL FREQUENCY WAS ZERO "
2560 PRINT "...............  ITS LOG WAS NOT DEFINED"
2570 GO TO 1460
2580 X1=P1
2590 X2=P1+D1
2600 Y1=P2
2610 Y2=Y1+D2
```

```
2620 VIEWPORT 20,120,28,98
2630 T1=(X2-X1)/100
2640 T2=(Y2-Y1)/100
2650 WINDOW X1-T1,X2+T1,Y1-T2,Y2+T2
2660 MOVE X1,Y1
2670 DRAW X2,Y1
2680 MOVE X1,Y1-8*T2
2690 PRINT "10"
2700 MOVE X1+2*T1,Y1-5*T2
2710 PRINT X1
2720 MOVE X1,Y1
2730 FOR I=X1 TO X2-1
2740 FOR J=1 TO 8
2750 MOVE I+LGT(J+1),Y1
2760 DRAW I+LGT(J+1),Y1+2*T2
2770 NEXT J
2780 MOVE I+1,Y1
2790 DRAW I+1,Y2
2800 MOVE I+1,Y1-8*T2
2810 PRINT "10"
2820 MOVE I+1+2*T1,Y1-5*T2
2830 PRINT I+1
2840 NEXT I
2850 MOVE X2-12*T1,Y1-8*T2
2860 PRINT "HERTZ"
2870 MOVE X1,Y1
2880 DRAW X1,Y2
2890 MOVE X1-8*T1,Y1
2900 PRINT "10"
2910 MOVE X1-6*T1,Y1+3*T2
2920 PRINT Y1
2930 MOVE X1,Y1
2940 FOR I=Y1 TO Y2-1
2950 FOR J=1 TO 8
2960 MOVE X1,I+LGT(J+1)
2970 DRAW X1+2*T1,I+LGT(J+1)
2980 NEXT J
2990 MOVE X1,I+1
3000 DRAW X2,I+1
3010 MOVE X1-8*T1,I+1
3020 PRINT "10"
3030 MOVE X1-6*T1,I+1+3*T2
3040 PRINT I+1
3050 NEXT I
3060 MOVE X1-8*T1,Y2-6*T2
3070 PRINT "AMP"
3080 GO TO 1790
3090 IF F>F1 THEN 3110
3100 MOVE LGT(F1),LGT(M1)
3110 DRAW LGT(F),LGT(M1)
3120 GO TO 2170
3130 REM AXIS SEMI-LOG
3140 PRINT
3150 PRINT
3160 PRINT "DEFINITION OF THE SCALE OF THE SEMI-LOG AXIS"
3170 PRINT "..............................................................................."
3180 PRINT
3190 PRINT "HOW MANY FREQUENCY DECADES DO YOU WISH   : ";
3200 INPUT D1
3210 PRINT
3220 PRINT "                                      N"
3230 PRINT "ORIGIN OF THE ABSCISSA  (10   HERTZ)    N= ";
```

```
3240 INPUT P1
3250 IF INT(P1)=P1 THEN 3300
3260 PRINT
3270 PRINT "CAUTION  N MUST BE AN INTEGER"
3280 PRINT "--------------------"
3290 GO TO 3210
3300 GO TO 1460
3310 PAGE
3320 IF F1<>0 THEN 3370
3330 PRINT "CAUTION  THE INITIAL FREQUENCY WAS ZERO"
3340 PRINT "--------------------"
3350 PRINT "            ITS LOG IS NOT DEFINED"
3360 GO TO 1460
3370 X1=F1
3380 X2=F1+D1
3390 Y1=0
3400 Y2=360
3410 VIEWPORT 20,120,28,98
3420 T1=(X2-X1)/100
3430 T2=3.6
3440 WINDOW X1-T1,X2+T1,Y1-T2,Y2+T2
3450 MOVE X1,Y1
3460 DRAW X2,Y1
3470 MOVE X1,Y1-8*T2
3480 PRINT "10"
3490 MOVE X1+2*T1,Y1-5*T2
3500 PRINT X1
3510 MOVE X1,Y1
3520 FOR I=X1 TO X2-1
3530 FOR J=1 TO 8
3540 MOVE I+LGT(J+1),Y1
3550 DRAW I+LGT(J+1),Y1+2*T2
3560 NEXT J
3570 MOVE I+1,Y1
3580 DRAW I+1,Y2
3590 MOVE I+1,Y1-8*T2
3600 PRINT "10"
3610 MOVE I+1+2*T1,Y1-5*T2
3620 PRINT I+1
3630 NEXT I
3640 MOVE X2-12*T1,Y1-8*T2
3650 PRINT "HERTZ"
3660 MOVE X1,Y1
3670 DRAW X1,Y2
3680 Y3=0
3690 MOVE X1-10*T1,Y1
3700 PRINT Y3
3710 FOR I=1 TO 4
3720 Y3=Y3+90
3730 MOVE X1-10*T1,Y3
3740 PRINT Y3
3750 MOVE X1,Y3
3760 DRAW X2,Y3
3770 NEXT I
3780 MOVE X1-10*T1,Y2-6*T2
3790 PRINT "PHASE"
3800 GO TO 1790
3810 IF F>F1 THEN 3830
3820 MOVE LGT(F1),T1
3830 DRAW LGT(F),T1
3840 GO TO 2170
3850 END
```

```
100 REM              PROGRAM NUMBER : 7
110 INIT
120 PAGE
130 PRINT "*********************************************************"
140 PRINT "*                                                       *"
150 PRINT "*                 STEADY STATE RESPONSE                 *"
160 PRINT "*    OF A SYSTEM SUBJECTED TO A SINUSOIDAL EXCITATION   *"
170 PRINT "*                         BY                            *"
180 PRINT "*                  THE MODAL METHOD                     *"
190 PRINT "*                                                       *"
200 PRINT "*                   (x)=[FI](q)                         *"
210 PRINT "*                                                       *"
220 PRINT "*              [FI]   MODAL MATRIX                      *"
230 PRINT "*          (q)    IMAGINARY PART OF (z)                 *"
240 PRINT "*                                                       *"
250 PRINT "*                        jWt                            *"
260 PRINT "*                  z=Ze                                 *"
270 PRINT "*                                                       *"
280 PRINT "*      (-W*W[M] + jW[C] + [K])(z)           jWt         *"
290 PRINT "*                   OR            = (Q1)+j(Q2)e         *"
300 PRINT "*      (-W*W[M] +   [K1]+j[K2])(z)                      *"
310 PRINT "*                                                       *"
320 PRINT "*         [M],[C],[K],[K1],[K2]  ARE DIAGONAL           *"
330 PRINT "*                                                       *"
340 PRINT "*      THE PHASE a IS DEFINED BY: x=!X! SIN(Wt-a)       *"
350 PRINT "*                                                       *"
360 PRINT "*********************************************************"
370 PRINT
380 PRINT
390 PRINT "NUMBER OF MODES = ";
400 INPUT N
410 DELETE M,K1,K2,Q1,Q2,G
420 N2=2*N
430 DIM M(N),K1(N),K2(N),Q1(N),Q2(N),G(N)
440 PRINT
450 PRINT
460 PRINT "MODAL MASSES [M]"
470 PRINT
480 FOR I=1 TO N
490 FOR J=I TO N
500 PRINT "M(";I;")= ";
510 INPUT M(I)
520 NEXT I
530 PRINT
540 PRINT "IF THE DAMPING IS VISCOUS      REF=VIS"
550 PRINT "            ''        STRUCTURAL   REF=STR"
560 PRINT "  REF= ";
570 INPUT C$
580 IF C$="VIS" THEN 630
590 IF C$="STR" THEN 610
600 GO TO 530
610 C9=2
620 GO TO 640
630 C9=1
640 PRINT
650 PRINT
660 GO TO C9 OF 670,690
670 PRINT "MODAL STIFFNESSES [K]"
680 GO TO 700
690 PRINT "MODAL STIFFNESSES [K1]"
700 PRINT
710 PRINT
720 FOR I=1 TO N
```

```
730 GO TO C9 OF 740,760
740 PRINT "K(";I;")= ";
750 GO TO 770
760 PRINT "K1(";I;")= ";
770 INPUT K1(I)
780 NEXT I
790 PRINT
800 PRINT
810 GO TO C9 OF 820,840
820 PRINT "MODAL DAMPING [C]"
830 GO TO 850
840 PRINT "MODAL DAMPING [K2]"
850 PRINT
860 PRINT
870 FOR I=1 TO N
880 GO TO C9 OF 890,910
890 PRINT "C(";I;")= ";
900 GO TO 920
910 PRINT "K2(";I;")= ";
920 INPUT K2(I)
930 NEXT I
940 PRINT
950 PRINT
960 PRINT "MODAL FORCES    (Q1) AND (Q2)"
970 PRINT
980 PRINT
990 FOR I=1 TO N
1000 PRINT "Q1(";I;")= ";
1010 INPUT Q1(I)
1020 NEXT I
1030 PRINT
1040 FOR I=1 TO N
1050 J=I-N
1060 PRINT "Q2(";I;")= ";
1070 INPUT Q2(I)
1080 NEXT I
1090 PRINT
1100 PRINT
1110 PRINT "NUMBER OF THE DEGREE OF FREEDOM TO BE CALCULATED NU= ";
1120 INPUT N3
1130 IF N3<1 OR N3>N THEN 1100
1140 PRINT
1150 PRINT "COMPONENTS OF THE MODES "
1160 PRINT "FOR THE DEGREE OF FREEDOM CONSIDERED"
1170 PRINT
1180 FOR I=1 TO N
1190 PRINT "FI(";N3;",";I;")= ";
1200 INPUT G(I)
1210 NEXT I
1220 PRINT
1230 PRINT "DO YOU WISH THE RESPONSE IN NUMERICAL FORM          REP=NUM"
1240 PRINT "            ''   TO PLOT THE AMPLITUDE OF THE RESPONSE  REP=AMP"
1250 PRINT "            ''   TO PLOT THE PHASE OF THE RESPONSE      REP=PHA"
1260 PRINT "            ''   TO STOP...                            REP=STOP"
1270 PRINT
1280 PRINT "  REP= ";
1290 INPUT A$
1300 IF A$="NUM" THEN 1350
1310 IF A$="AMP" THEN 1370
1320 IF A$="PHA" THEN 1390
1330 IF A$="STOP" THEN 3810
1340 GO TO 1220
1350 C1=1
```

```
1360 GO TO 1430
1370 C1=2
1380 GO TO 1430
1390 C1=3
1400 GO TO 1430
1410 C1=4
1420 GO TO 1140
1430 GO TO C1 OF 1440,2250,3100,3810
1440 PRINT
1450 PRINT
1460 PRINT "SWEEPING"
1470 PRINT "—————————"
1480 PRINT
1490 PRINT "INITIAL FREQUENCY OF THE SWEEP  (HERTZ) = ";
1500 INPUT F1
1510 PRINT "FINAL              ''              ''    = ";
1520 INPUT F2
1530 PRINT
1540 PRINT "DO YOU WISH A LINEAR        INCREMENT   REP=LIN"
1550 PRINT "        ''        LOGARITHMIC        ''        REP=LOG"
1560 PRINT "  REP= ";
1570 INPUT B$
1580 PRINT
1590 IF B$="LIN" THEN 1620
1600 IF B$="LOG" THEN 1650
1610 GO TO 1530
1620 PRINT "INCREMENT            ''              ''    = ";
1630 INPUT F3
1640 GO TO 1750
1650 IF F1<>0 THEN 1700
1660 PRINT
1670 PRINT "CAUTION  THE INITIAL FREQUENCY WAS ZERO"
1680 PRINT "————————  ITS LOG IS NOT DEFINED"
1690 GO TO 1440
1700 PRINT
1710 PRINT "NUMBER OF EQUAL DIVISIONS ALONG"
1720 PRINT "THE LOGARITHMIC SCALE BETWEEN FI AND F2 = ";
1730 INPUT F3
1740 F3=10^((LGT(F2)-LGT(F1))/F3)
1750 GO TO C1 OF 1760,2500,3280,3810
1760 PAGE
1770 F=F1
1780 W=2*PI*F
1790 X1=0
1800 X2=0
1810 FOR I=1 TO N
1820 GO TO C9 OF 1830,1870
1830 D=(K1(I)-M(I)*W*W)^2+K2(I)^2*W^2
1840 R1=Q1(I)*(K1(I)-M(I)*W*W)+Q2(I)*K2(I)*W
1850 R2=Q2(I)*(K1(I)-M(I)*W*W)-Q1(I)*K2(I)*W
1860 GO TO 1900
1870 D=(K1(I)-M(I)*W*W)^2+K2(I)^2
1880 R1=Q1(I)*(K1(I)-M(I)*W*W)+Q2(I)*K2(I)
1890 R2=Q2(I)*(K1(I)-M(I)*W*W)-Q1(I)*K2(I)
1900 X1=X1+R1/D*G(I)
1910 X2=X2+R2/D*G(I)
1920 NEXT I
1930 M1=SQR(X1^2+X2^2)
1940 IF M1<>0 THEN 1970
1950 T1=0
1960 GO TO 2100
1970 O1=X1/M1
1980 O1=O1 O1/1 OE+9
```

```
1990 I1=-X2/M1
2000 IF O1<>-1 THEN 2030
2010 T1=+PI
2020 GO TO 2100
2030 IF O1<>1 THEN 2060
2040 T1=0
2050 GO TO 2100
2060 IF I1<0 THEN 2090
2070 T1=ACS(O1)
2080 GO TO 2100
2090 T1=2*PI-ACS(O1)
2100 T1=T1/PI*180
2110 GO TO C1 OF 2120,3060,3770,3810
2120 PRINT USING 2130:"FREQUENCY= ",F,"! DISP!= ",M1," PHASE=",T1," DEG"
2130 IMAGE 11A,3E,9A,3E,8A,4D.2D,4A
2140 IF B$="LIN" THEN 2170
2150 F=F*F3
2160 GO TO 2180
2170 F=F+F3
2180 IF F>F2 THEN 2210
2190 GO TO 1780
2200 NEXT F
2210 PRINT
2220 GO TO C1 OF 1220,2230,2230,3810
2230 MOVE X1,Y1-10*T2
2240 GO TO 1220
2250 REM AXIS LOG-LOG
2260 PRINT
2270 PRINT
2280 PRINT "DEFINITION OF THE SCALE OF THE LOG-LOG AXIS"
2290 PRINT "..................................................................................................."
2300 PRINT
2310 PRINT "HOW MANY FREQUENCY DECADES DO YOU WISH   : ";
2320 INPUT D1
2330 PRINT "      ''     RESPONSE              ''                : ";
2340 INPUT D2
2350 PRINT
2360 PRINT "                              N"
2370 PRINT "ORIGIN OF THE ABSCISSA  (10   HERTZ)     N= ";
2380 INPUT P1
2390 IF INT(P1)<>P1 THEN 2450
2400 PRINT
2410 PRINT "                              N"
2420 PRINT "      ''         ORDINATE   (10   MKSA )     N= ";
2430 INPUT P2
2440 IF INT(P2)=P2 THEN 2490
2450 PRINT
2460 PRINT "CAUTION  N MUST BE AN INTEGER"
2470 PRINT "..............."
2480 GO TO 2260
2490 GO TO 1440
2500 PAGE
2510 IF F1<>0 THEN 2550
2520 PRINT "CAUTION   THE INITIAL FREQUENCY WAS ZERO"
2530 PRINT "..............   ITS LOG IS NOT DEFINED"
2540 GO TO 1440
2550 X1=P1
2560 X2=P1+D1
2570 Y1=P2
2580 Y2=Y1+D2
2590 VIEWPORT 20,120,28,98
2600 T1=(X2-X1)/100
2610 T2=(Y2-Y1)/100
```

```
2620 WINDOW X1-T1,X2+T1,Y1-T2,Y2+T2
2630 MOVE X1,Y1
2640 DRAW X2,Y1
2650 MOVE X1,Y1-8*T2
2660 PRINT "10"
2670 MOVE X1+2*T1,Y1-5*T2
2680 PRINT X1
2690 MOVE X1,Y1
2700 FOR I=X1 TO X2-1
2710 FOR J=1 TO 8
2720 MOVE I+LGT(J+1),Y1
2730 DRAW I+LGT(J+1),Y1+2*T2
2740 NEXT J
2750 MOVE I+1,Y1
2760 DRAW I+1,Y2
2770 MOVE I+1,Y1-8*T2
2780 PRINT "10"
2790 MOVE I+1+2*T1,Y1-5*T2
2800 PRINT I+1
2810 NEXT I
2820 MOVE X2-12*T1,Y1-8*T2
2830 PRINT "HERTZ"
2840 MOVE X1,Y1
2850 DRAW X1,Y2
2860 MOVE X1-8*T1,Y1
2870 PRINT "10"
2880 MOVE X1-6*T1,Y1+3*T2
2890 PRINT Y1
2900 MOVE X1,Y1
2910 FOR I=Y1 TO Y2-1
2920 FOR J=1 TO 8
2930 MOVE X1,I+LGT(J+1)
2940 DRAW X1+2*T1,I+LGT(J+1)
2950 NEXT J
2960 MOVE X1,I+1
2970 DRAW X2,I+1
2980 MOVE X1-8*T1,I+1
2990 PRINT "10"
3000 MOVE X1-6*T1,I+1+3*T2
3010 PRINT I+1
3020 NEXT I
3030 MOVE X1-8*T1,Y2-6*T2
3040 PRINT "AMP"
3050 GO TO 1770
3060 IF F>F1 THEN 3080
3070 MOVE LGT(F1),LGT(M1)
3080 DRAW LGT(F),LGT(M1)
3090 GO TO 2140
3100 REM AXIS SEMI-LOG
3110 PRINT
3120 PRINT
3130 PRINT "DEFINITION OF THE SCALE OF THE SEMI-LOG AXIS"
3140 PRINT "---------------------------------------------------------------------"
3150 PRINT
3160 PRINT "HOW MANY FREQUENCY DECADES DO YOU WISH  : ";
3170 INPUT D1
3180 PRINT
3190 PRINT "                                    N"
3200 PRINT "ORIGIN OF THE ABSCISSA  (10  HERTZ)    N= ";
3210 INPUT P1
3220 IF INT(P1)=P1 THEN 3270
3230 PRINT
3240 PRINT "CAUTION  N MUST BE AN INTEGER"
```

```
3250 PRINT "-----------------"
3260 GO TO 3180
3270 GO TO 1440
3280 PAGE
3290 IF F1<>0 THEN 3330
3300 PRINT "CAUTION   THE INITIAL FREQUENCY WAS ZERO"
3310 PRINT "-----------   ITS LOG IS NOT DEFINED"
3320 GO TO 1440
3330 X1=P1
3340 X2=P1+D1
3350 Y1=0
3360 Y2=360
3370 VIEWPORT 20,120,28,98
3380 T1=(X2-X1)/100
3390 T2=3.6
3400 WINDOW X1-T1,X2+T1,Y1-T2,Y2+T2
3410 MOVE X1,Y1
3420 DRAW X2,Y1
3430 MOVE X1,Y1-8*T2
3440 PRINT "10"
3450 MOVE X1+2*T1,Y1-5*T2
3460 PRINT X1
3470 MOVE X1,Y1
3480 FOR I=X1 TO X2-1
3490 FOR J=1 TO 8
3500 MOVE I+LGT(J+1),Y1
3510 DRAW I+LGT(J+1),Y1+2*T2
3520 NEXT J
3530 MOVE I+1,Y1
3540 DRAW I+1,Y2
3550 MOVE I+1,Y1-8*T2
3560 PRINT "10"
3570 MOVE I+1+2*T1,Y1-5*T2
3580 PRINT I+1
3590 NEXT I
3600 MOVE X2-12*T1,Y1-8*T2
3610 PRINT "HERTZ"
3620 MOVE X1,Y1
3630 DRAW X1,Y2
3640 Y3=0
3650 MOVE X1-10*T1,Y1
3660 PRINT Y3
3670 FOR I=1 TO 4
3680 Y3=Y3+90
3690 MOVE X1-10*T1,Y3
3700 PRINT Y3
3710 MOVE X1,Y3
3720 DRAW X2,Y3
3730 NEXT I
3740 MOVE X1-10*T1,Y2-6*T2
3750 PRINT "PHASE"
3760 GO TO 1770
3770 IF F>F1 THEN 3790
3780 MOVE LGT(F1),T1
3790 DRAW LGT(F),T1
3800 GO TO 2140
3810 END


100 REM          PROGRAM NUMBER : 8
110 INIT
120 PAGE
130 PRINT
```

```
140 PRINT "****************************************"
150 PRINT "*                                      *"
160 PRINT "*          RESPONSE OF A SYSTEM        *"
170 PRINT "*      TO AN ARBITRARY FORCING FUNCTION *"
180 PRINT "*                                      *"
190 PRINT "*     [M](x)''+[C](x)''+[K](x)={f(t)}  *"
200 PRINT "*                                      *"
210 PRINT "*                 BY                   *"
220 PRINT "*        THE WILSON-THETA METHOD       *"
230 PRINT "*                                      *"
240 PRINT "****************************************"
250 PRINT
260 PRINT
270 PRINT "NUMBER OF DEGREES OF FREEDOM= ";
280 INPUT N
290 DELETE M,C,K,X0,V0,G0,F0,K1,F1,F2,X1,V1,G1,X2
300 DIM M(N,N),C(N,N),K(N,N),X0(N),V0(N),G0(N),F0(N),K1(N,N),F1(N),F2(N
310 DIM X1(N),V1(N),G1(N),X2(N)
320 REM
330 PRINT
340 PRINT "MASS MATRIX [M]"
350 PRINT
360 FOR I=1 TO N
370 FOR J=I TO N
380 PRINT "M(";I;",";J;")= ";
390 INPUT M(I,J)
400 NEXT J
410 NEXT I
420 PRINT
430 PRINT "DAMPING MATRIX [C]"
440 PRINT
450 FOR I=1 TO N
460 FOR J=I TO N
470 PRINT "C(";I;",";J;")= ";
480 INPUT C(I,J)
490 NEXT J
500 NEXT I
510 PRINT
520 PRINT "STIFFNESS MATRIX [K]"
530 PRINT
540 FOR I=1 TO N
550 FOR J=I TO N
560 PRINT "K(";I;",";J;")= ";
570 INPUT K(I,J)
580 NEXT J
590 NEXT I
600 FOR I=1 TO N
610 FOR J=I TO N
620 K(J,I)=K(I,J)
630 C(J,I)=C(I,J)
640 M(J,I)=M(I,J)
650 NEXT J
660 NEXT I
670 PRINT
680 PRINT "INITIAL    TIME    (SEC) TO = ";
690 INPUT T0
700 PRINT "FINAL      ''      (SEC) TF = ";
710 INPUT T3
720 PRINT "INCREMENT  ''      (SEC)INC = ";
730 INPUT D
740 PRINT
750 FOR I=1 TO N
760 PRINT "DISPLACEMENT X(";I;") AT TO = ";
```

```
770 INPUT X0(I)
780 NEXT I
790 PRINT
800 FOR I=1 TO N
810 PRINT "VELOCITY      V(";I;") AT TO = ";
820 INPUT V0(I)
830 NEXT I
840 PRINT
850 FOR I=1 TO N
860 PRINT "ACCELERATION G(";I;") AT TO = ";
870 INPUT G0(I)
880 NEXT I
890 PRINT
900 PRINT "VALUE OF THETA= ";
910 INPUT T2
920 PRINT
930 PRINT "NUMBER OF FORCE COMPONENT= ";
940 INPUT N1
950 IF N1<1 OR N1>N THEN 920
960 PRINT
970 PRINT "AT THE TIME t= ";TO;" THE FORCE HAS THE VALUE = ";
980 INPUT F
990 PRINT
1000 F0=0
1010 F0(N1)=F
1020 PAGE
1030 A9=T2*D
1040 A0=6/A9^2
1050 A1=3/A9
1060 A2=2*A1
1070 A3=A9/2
1080 A4=A0/T2
1090 A5=-A2/T2
1100 A6=1-3/T2
1110 A7=D/2
1120 A8=D^2/6
1130 REM
1140 REM * FORMATION OF K+A0*M+A1*C *
1150 REM      --------------------------------
1160 F1=0
1170 T1=TO+D
1180 PRINT USING 1190:"AT THE TIME t= ",T1," THE FORCE HAS THE VALUE= ";
1190 IMAGE 15A,3E,27A,S
1200 INPUT F
1210 PRINT "--------------------------------------------------------------------------------"
1220 F1(N1)=F
1230 F2=0
1240 FOR I=1 TO N
1250 F2(I)=+F0(I)+T2*(F1(I)-F0(I))
1260 FOR J=1 TO N
1270 K1(I,J)=K(I,J)+A0*M(I,J)+A1*C(I,J)
1280 F2(I)=F2(I)+M(I,J)*(A0*X0(J)+A2*V0(J)+2*G0(J))
1290 F2(I)=F2(I)+C(I,J)*(A1*X0(J)+2*V0(J)+A3*G0(J))
1300 NEXT J
1310 NEXT I
1320 K1=INV(K1)
1330 X2=K1 MPY F2
1340 FOR I=1 TO N
1350 G1(I)=A4*(X2(I)-X0(I))+A5*V0(I)+A6*G0(I)
1360 V1(I)=V0(I)+A7*(G1(I)+G0(I))
1370 X1(I)=X0(I)+D*V0(I)+A8*(G1(I)+2*G0(I))
1380 NEXT I
1390 PRINT
```

```
1400 PRINT USING 1410:"DISPLACEMENT","VELOCITY","ACCELERATION"
1410 IMAGE 5X,12A,12X,8A,15X,12A
1420 PRINT
1430 FOR I=1 TO N
1440 PRINT USING 1450:"X(",I,")= ",X1(I),"    V(",I,")= ",V1(I)
1450 IMAGE 2A,2D,3A,3E,7A,2D,3A,3E,S
1460 PRINT USING 1470:"    G(",I,")= ",G1(I)
1470 IMAGE 7A,2D,3A,3E
1480 NEXT I
1490 PRINT
1500 PRINT
1510 X0=X1
1520 V0=V1
1530 G0=G1
1540 T0=T1
1550 F0=F1
1560 IF T1<T3 THEN 1160
1570 PRINT
1580 PRINT "DO YOU WISH TO REDO THE SWEEP                  REP=SWE"
1590 PRINT "              ''       CHANGE THE INITIAL CONDITIONS  REP=INI"
1600 PRINT "              ''       CHANGE THE SYSTEM              REP=SYS"
1610 PRINT "              ''       STOP...                        REP=STOP"
1620 PRINT
1630 PRINT "  REP= ";
1640 INPUT A$
1650 IF A$="SWE" THEN 670
1660 IF A$="INI" THEN 740
1670 IF A$="SYS" THEN 260
1680 IF A$="STOP" THEN 1700
1690 GO TO 1570
1700 END


100 REM            PROGRAM NUMBER : 9
110 INIT
120 PAGE
130 PRINT "***********************************************"
140 PRINT "*                                             *"
150 PRINT "*                 FREQUENCIES                 *"
160 PRINT "*            OF A BEAM IN BENDING             *"
170 PRINT "*         MODAL MASSES AND STIFFNESSES        *"
180 PRINT "*    MODES:DEFLECTION,SLOPE,MAXIMUM STRESS    *"
190 PRINT "*                                             *"
200 PRINT "***********************************************"
210 PRINT
220 PRINT
230 PRINT "IF THE BEAM IS C-F    REP=1"
240 PRINT "              ''      C-S    REP=2"
250 PRINT "              ''      C-C    REP=3"
260 PRINT "              ''      F-F    REP=4"
270 PRINT "              ''      F-S    REP=5"
280 PRINT "              ''      S-S    REP=6"
290 PRINT
300 PRINT "  REP = ";
310 INPUT M
320 IF M<1 OR M>6 THEN 220
330 DELETE A1
340 DIM A1(6)
350 GO TO M OF 360,570,500,500,570,430
360 REM C-F
370 A1(1)=3.516
380 A1(2)=22.034
390 A1(3)=61.697
```

```
400 A1(4)=120.9
410 A1(5)=199.86
420 GO TO 630
430 REM S-S
440 A1(1)=9.8696
450 A1(2)=39.478
460 A1(3)=88.826
470 A1(4)=157.91
480 A1(5)=246.74
490 GO TO 630
500 REM  C-C AND F-F
510 A1(1)=22.373
520 A1(2)=61.673
530 A1(3)=120.9
540 A1(4)=199.86
550 A1(5)=298.56
560 GO TO 630
570 REM  C-S AND F-S
580 A1(1)=15.418
590 A1(2)=49.965
600 A1(3)=104.25
610 A1(4)=178.27
620 A1(5)=272.03
630 PRINT
640 PRINT "YOUNG'S MODULUS......................= ";
650 INPUT E
660 PRINT "MASS DENSITY      ...................= ";
670 INPUT R
680 PRINT "CROSS SECTIONAL AREA.................= ";
690 INPUT S
700 PRINT "MOMENT OF INERTIA OF THE CROSS SECTION= ";
710 INPUT J
720 PRINT "LENGTH...............................= ";
730 INPUT L
740 PRINT "DISTANCE OF THE FURTHEST POINT.......= ";
750 INPUT DO
760 PRINT
770 PRINT
780 FOR I=1 TO 5
790 B1=SQR(A1(I))
800 F=A1(I)/(2*PI*L^2)*SQR(E*J/R/S)
810 PRINT USING 820: "FREQUENCY = ",F," HERTZ"
820 IMAGE 12A,3E,6A
830 NEXT I
840 PRINT
850 PRINT
860 PRINT "DO YOU WISH THE MODES   YES OR NO : ";
870 INPUT A$
880 IF A$="YES" THEN 910
890 IF A$="NO" THEN 2110
900 GO TO 850
910 PRINT
920 PRINT "NUMBER OF MODES (NOT MORE THAN FIVE)= ";
930 INPUT N1
940 PRINT
950 PRINT "AT HOW MANY POINTS DO YOU WISH THE DEFLECTION,SLOPE,"
960 PRINT "AND THE MAXIMUM STRESS: ";
970 INPUT N2
980 DELETE X
990 DIM X(N2)
1000 PRINT
1010 PRINT "DO YOU WISH AN AUTOMATIC DEFINITION OF THE INTERVAL"
1020 PRINT "  YES OR NO : ";
```

```
1030 INPUT A$
1040 IF A$="YES" THEN 1130
1050 IF A$="NO" THEN 1070
1060 GO TO 1000
1070 PRINT
1080 FOR I=1 TO N2
1090 PRINT "X(";I;")= ";
1100 INPUT X(I)
1110 NEXT I
1120 GO TO 1180
1130 L0=L/(N2-1)
1140 X(1)=0
1150 FOR I=2 TO N2
1160 X(I)=X(I-1)+L0
1170 NEXT I
1180 PAGE
1190 FOR I=1 TO N1
1200 B1=SQR(A1(I))
1210 B2=B1/L
1220 O1=COS(B1)
1230 I1=SIN(B1)
1240 O2=COS(2*B1)
1250 I2=SIN(2*B1)
1260 C1=(EXP(B1)+EXP(-B1))/2
1270 S1=(EXP(B1)-EXP(-B1))/2
1280 C2=(EXP(2*B1)+EXP(-2*B1))/2
1290 S2=(EXP(2*B1)-EXP(-2*B1))/2
1300 A=1
1310 GO TO M OF 1320,1370,1370,1420,1470,1520
1320 REM C-F
1330 B=-(I1+S1)/(O1+C1)
1340 C=-1
1350 D=-B
1360 GO TO 1560
1370 REM C-S AND C-C
1380 B=(S1-I1)/(O1-C1)
1390 C=-1
1400 D=-B
1410 GO TO 1560
1420 REM F-F
1430 B=(I1-S1)/(C1-O1)
1440 C=1
1450 D=B
1460 GO TO 1560
1470 REM F-S
1480 B=-(I1+S1)/(O1+C1)
1490 C=1
1500 D=B
1510 GO TO 1560
1520 REM S-S
1530 B=0
1540 C=0
1550 D=0
1560 T1=A^2/2*(L-1/2/B2*I2)
1570 T2=B^2/2*(L+I2/2/B2)
1580 T3=C^2/2*(S2/2/B2-L)
1590 T4=D^2/2*(S2/2/B2+L)
1600 T5=A*B/2/2/B2*(1-O2)
1610 T6=C*D/2/2/B2*(C2-1)
1620 T7=EXP(+B1)/2/2/B2*(A*C+A*D)*(I1-O1)
1630 T7=T7-EXP(-B1)/2/2/B2*(A*D-A*C)*(I1+O1)
1640 T7=T7+A*D/2/B2
1650 T8=EXP(B1)/2/2/B2*(B*C+B*D)*(I1+O1)
```

```
1660 T8=T8+EXP(-B1)/2/2/B2*(B*D-B*C)*(-O1+I1)
1670 T8=T8-B*C/2/B2
1680 T=T1+T2+T3+T4+2*T5+2*T6+2*T7+2*T8
1690 M1=R*S*T
1700 K1=M1*B2^4*E*J/R/S
1710 W=SQR(K1/M1)
1720 F=W/2/PI
1730 PRINT
1740 PRINT "          MODE NUMBER : ";I
1750 PRINT "          ............................."
1760 PRINT
1770 PRINT USING 1780:"FREQUENCY [RAD/SEC]= ",W," FREQUENCY [HERTZ]= ",F
1780 IMAGE21A,3E,20A,3E
1790 PRINT
1800 PRINT USING 1810:"MODAL MASS = ",M1,"MODAL STIFFNESS= ",K1
1810 IMAGE 13A,3E,9X,18A,3E
1820 PRINT
1830 PRINT
1840 D$="***********"
1850 PRINT D$;D$;D$;D$;D$
1860 PRINT USING 1870:"*","*","*","*","*"
1870 IMAGE 1A,16X,1A,15X,1A,15X,1A,15X,1A
1880 PRI USI 1890:"*","X","*","DEFLECTION","*","SLOPE","*","STRESS","*"
1890 IMAGE 1A,7X,1A,8X,1A,2X,10A,3X,1A,5X,5A,5X,1A,4X,6A,5X,1A
1900 PRINT USING 1870:"*","*","*","*","*"
1910 PRINT D$;D$;D$;D$;D$
1920 PRINT USING 1870:"*","*","*","*","*"
1930 FOR K=1 TO N2
1940 Z=X(K)
1950 I1=SIN(B2*Z)
1960 O1=COS(B2*Z)
1970 S1=(EXP(B2*Z)-EXP(-B2*Z))/2
1980 C1=(EXP(B2*Z)+EXP(-B2*Z))/2
1990 Y=A*I1+B*O1+C*S1+D*C1
2000 P=B2*(A*O1-B*I1+C*C1+D*S1)
2010 M1=B2^2*(-A*I1-B*O1+C*S1+D*C1)
2020 M1=-D0*E*M1
2030 PRINT USING 2040:"*",Z,"*",Y,"*",P,"*",M1,"*"
2040 IMAGE 1A,2X,4E,2X,3(1A,2X,3E,2X),1A
2050 NEXT K
2060 PRINT USING 1870:"*","*","*","*","*"
2070 PRINT D$;D$;D$;D$;D$;D$
2080 PRINT
2090 PRINT
2100 NEXT I
2110 END

100 REM           PROGRAM NUMBER :10
110 INIT
120 PAGE
130 PRINT "***********************************"
140 PRINT "*                                 *"
150 PRINT "*           FREQUENCIES           *"
160 PRINT "*     OF A SYSTEM COMPOSED OF      *"
170 PRINT "*        -BEAMS                    *"
180 PRINT "*        -SUPPORTED INERTIAS       *"
190 PRINT "*        -SUSPENDED INERTIAS       *"
200 PRINT "*               BY                 *"
210 PRINT "*    THE TRANSFER MATRIX METHOD    *"
220 PRINT "*                                 *"
230 PRINT "***********************************"
240 PRINT
250 PRINT "TYPE OF ELEMENT"
```

```
260 PRINT "——————————————————————"
270 PRINT "BEAM                    REP=B"
280 PRINT "SUPPORTED INERTIA   REP=SUP"
290 PRINT "SUSPENDED    ''      REP=SUS"
300 PRINT
310 PRINT "NUMBER OF ELEMENTS= ";
320 INPUT N
330 P9=0
340 R9=0
350 S9=0
360 DELETE C,T,T1,A
370 DIM C(N),T(4,4),T1(4,4),A(4,4)
380 PRINT
390 FOR I=1 TO N
400 PRINT "TYPE OF ELEMENT ";I;"  REP= ";
410 INPUT C$
420 IF C$="B" THEN 460
430 IF C$="SUP" THEN 490
440 IF C$="SUS" THEN 520
450 GO TO 240
460 C(I)=1
470 P9=P9+1
480 GO TO 540
490 C(I)=2
500 R9=R9+1
510 GO TO 540
520 C(I)=3
530 S9=S9+1
540 NEXT I
550 IF P9=0 THEN 720
560 PRINT
570 DELETE E,R,L1,S5,I5
580 DIM E(P9),R(P9),L1(P9),S5(P9),I5(P9)
590 FOR I=1 TO P9
600 PRINT "BEAM NU:";I;" YOUNG'S MODULUS.....................= ";
610 INPUT E(I)
620 PRINT "BEAM NU:";I;" MASS DENSITY.......................= ";
630 INPUT R(I)
640 PRINT "BEAM NU:";I;" CROSS SECTIONAL AREA...............= ";
650 INPUT S5(I)
660 PRINT "BEAM NU:";I;" MOMENT OF INERTIA OF THE CROSS SECTION= ";
670 INPUT I5(I)
680 PRINT "BEAM NU:";I;" LENGTH.............................= ";
690 INPUT L1(I)
700 PRINT
710 NEXT I
720 DELETE M1,I1,K1,R1,M2,I2,K2,R2
730 IF R9=0 THEN 890
740 PRINT
750 DIM M1(R9),I1(R9),K1(R9),R1(R9)
760 FOR I=1 TO R9
770 PRINT "FOR THE SUPPORTED INERTIA NU: ";I
780 PRINT
790 PRINT "MASS............= ";
800 INPUT M1(I)
810 PRINT "ROTARY INERTIA..= ";
820 INPUT I1(I)
830 PRINT "STIFFNESS.......= ";
840 INPUT K1(I)
850 PRINT "ROTARY STIFFNESS= ";
860 INPUT R1(I)
870 PRINT
880 NEXT I
```

```
 890 IF S9=0 THEN 1050
 900 PRINT
 910 DIM M2(S9),I2(S9),K2(S9),R2(S9)
 920 FOR I=1 TO S9
 930 PRINT "FOR THE SUSPENDED INERTIA NU: ";I
 940 PRINT
 950 PRINT "MASS............= ";
 960 INPUT M2(I)
 970 PRINT "ROTARY INERTIA..= ";
 980 INPUT I2(I)
 990 PRINT "STIFFNESS.......= ";
1000 INPUT K2(I)
1010 PRINT "ROTARY STIFFNESS= ";
1020 INPUT R2(I)
1030 PRINT
1040 NEXT I
1050 PRINT
1060 PRINT "IF THE SYSTEM IS C-C   REP=1"
1070 PRINT "                 ''    C-F   REP=2"
1080 PRINT "                 ''    C-S   REP=3"
1090 PRINT "                 ''    F-F   REP=4"
1100 PRINT "                 ''    F-S   REP=5"
1110 PRINT "                 ''    S-S   REP=6"
1120 PRINT "   REP= ";
1130 INPUT C8
1140 PRINT
1150 PRINT "SWEEPING"
1160 PRINT "--------------"
1170 PRINT
1180 PRINT "INITIAL FREQUENCY OF THE SWEEP IN  (HERTZ) = ";
1190 INPUT F1
1200 IF F1<>0 THEN 1250
1210 PRINT
1220 PRINT "CAUTION   THE EXACT TRANSFER MATRIX IS NOT DEFINED"
1230 PRINT "--------   IF THE FREQUENCY IS ZERO"
1240 GO TO 1140
1250 PRINT "FINAL       ''                         ''   = ";
1260 INPUT F2
1270 PRINT "INCREMENT   ''                         ''   = ";
1280 INPUT P
1290 PAGE
1300 FOR F=F1 TO F2 STEP P
1310 P9=0
1320 R9=0
1330 S9=0
1340 W=2*PI*F
1350 W2=W^2
1360 T=0
1370 FOR I=1 TO N
1380 C2=C(I)
1390 GO TO C2 OF 1750,2080,2140
1400 REM
1410 IF I>1 THEN 1440
1420 T=A
1430 GO TO 1460
1440 T1=A MPY T
1450 T=T1
1460 NEXT I
1470 GO TO C8 OF 1480,1500,1520,1540,1560,1580
1480 D=T(3,1)*T(4,2)-T(4,1)*T(3,2)
1490 GO TO 1590
1500 D=T(1,1)*T(2,2)-T(2,1)*T(1,2)
1510 GO TO 1590
```

```
1520 D=T(2,1)*T(4,2)-T(4,1)*T(2,2)
1530 GO TO 1590
1540 D=T(1,3)*T(2,4)-T(2,3)*T(1,4)
1550 GO TO 1590
1560 D=T(2,3)*T(4,4)-T(4,3)*T(2,4)
1570 GO TO 1590
1580 D=T(2,1)*T(4,3)-T(4,1)*T(2,3)
1590 PRINT USING 1600:"FREQUENCY IN HERTZ = ",F,"    DETERMINANT = ",D
1600 IMAGE 21A,5E,18A,3E
1610 NEXT F
1620 PRINT
1630 PRINT
1640 PRINT "DO YOU WISH TO REDO THE SWEEP                      REP=SWE"
1650 PRINT "              ''       CHANGE THE BOUNDARY CONDITIONS  REP=BOU"
1660 PRINT "              ''       CHANGE THE SYSTEM              REP=SYS"
1670 PRINT "              ''       STOP                          REP=STOP"
1680 PRINT "  REP= ";
1690 INPUT A$
1700 IF A$="SWE" THEN 1140
1710 IF A$="BOU" THEN 1050
1720 IF A$="SYS" THEN 240
1730 IF A$="STOP" THEN 2260
1740 GO TO 1630
1750 REM TRANSFER MATRIX OF THE BEAM
1760 P9=P9+1
1770 E0=E(P9)
1780 S6=S5(P9)
1790 L3=L1(P9)
1800 I6=I5(P9)
1810 R0=R(P9)
1820 B2=W2*S6*R0/(I6*E0)
1830 B=SQR(B2)
1840 B0=SQR(B)
1850 E1=E0*I6
1860 B4=B0*L3
1870 C1=(EXP(B4)+EXP(-B4))/2
1880 O1=COS(B4)
1890 S1=(EXP(B4)-EXP(-B4))/2
1900 N1=SIN(B4)
1910 A(1,1)=(C1+O1)/2
1920 A(1,2)=B0*(S1-N1)/2
1930 A(1,3)=E1*B*(C1-O1)/2
1940 A(1,4)=E1*B*B0*(S1+N1)/2
1950 A(2,1)=(S1+N1)/(2*B0)
1960 A(2,2)=A(1,1)
1970 A(2,3)=E1*B0*(S1-N1)/2
1980 A(2,4)=A(1,3)
1990 A(3,1)=(C1-O1)/(2*E1*B)
2000 A(3,2)=(S1+N1)/(2*E1*B0)
2010 A(3,3)=A(1,1)
2020 A(3,4)=B0*(S1-N1)/2
2030 A(4,1)=(S1-N1)/(2*E1*B*B0)
2040 A(4,2)=A(3,1)
2050 A(4,3)=A(2,1)
2060 A(4,4)=A(1,1)
2070 GO TO 1400
2080 REM TRANSFER MATRIX SUP
2090 R9=R9+1
2100 CALL "IDN",A
2110 A(1,4)=M1(R9)*W2-K1(R9)
2120 A(2,3)=R1(R9)-I1(R9)*W2
2130 GO TO 1400
2140 REM TRANSFER MATRIX SUS
```

```
2150 S9=S9+1
2160 CALL "IDN",A
2170 IF K2(S9)=0 AND M2(S9)=0 THEN 2200
2180 A(1,4)=+K2(S9)*M2(S9)*W2/(K2(S9)-M2(S9)*W2)
2190 GO TO 2210
2200 A(1,4)=0
2210 IF R2(S9)=0 AND I2(S9)=0 THEN 2240
2220 A(2,3)=-R2(S9)*I2(S9)*W2/(R2(S9)-I2(S9)*W2)
2230 GO TO 2250
2240 A(2,3)=0
2250 GO TO 1400
2260 END


100 REM               PROGRAM NUMBER :11  * FIRST PART *
110 INIT
120 PAGE
130 PRINT "************************************************"
140 PRINT "*                                             *"
150 PRINT "*    FREQUENCIES,MODES AND ELEMENT ENERGIES   *"
160 PRINT "*    OF A PLANE FRAME COMPOSED OF             *"
170 PRINT "*        -   BEAMS AND BARS ELEMENTS          *"
180 PRINT "*        -   MASS-SPRING-MASS SYSTEMS         *"
190 PRINT "*        -   ARBITRARY ELEMENTS              *"
200 PRINT "*                    BY                       *"
210 PRINT "*          THE FINITE ELEMENT METHOD          *"
220 PRINT "*                                             *"
230 PRINT "************************************************"
240 PRINT
250 PRINT
260 PRINT " NUMBER OF ELEMENTS........ = ";
270 INPUT N7
280 PRINT "      ''      NODES........... = ";
290 INPUT N6
300 PRINT "      ''      MATERIALS....... = ";
310 INPUT N9
320 PRINT "      ''      SECTIONS........ = ";
330 INPUT N8
340 PRINT "      ''      CONSTRAINED NODES= ";
350 INPUT N5
360 PRINT
370 N=N6*3
380 DELETE NO
390 DIM NO(N7)
400 PRINT
410 PRINT "TYPE OF ELEMENT"
420 PRINT "----------------------------------------"
430 PRINT "BEAM AND BAR        REF=B"
440 PRINT "MASS-SPRING-MASS    REF=MSM"
450 PRINT "ARBITRARY           REF=ARB"
460 PRINT
470 P9=0
480 R9=0
490 Q9=0
500 FOR I=1 TO N7
510 PRINT "TYPE OF ELEMENT NU: ";I;" REF = ";
520 INPUT A$
530 IF A$="B" THEN 570
540 IF A$="MSM" THEN 600
550 IF A$="ARB" THEN 630
560 GO TO 510
570 P9=P9+1
580 NO(I)=1
590 GO TO 690
```

```
600 R9=R9+1
610 NO(I)=2
620 GO TO 690
630 Q9=Q9+1
640 NO(I)=3
650 IF Q9<=1 THEN 690
660 PRINT "CAUTION  ONLY ONE ARBITRARY ELEMENT"
670 PRINT "------------------"
680 GO TO 400
690 NEXT I
700 DELETE N1,N2
710 DIM N1(N7),N2(N7)
720 IF P9=0 THEN 1020
730 DELETE M3,S2,E1,R1,S1,I1
740 DIM M3(N7),S2(N7)
750 PRINT
760 FOR I=1 TO P9
770 PRINT "BEAM ELEMENT NU: ";I;" NUMBER OF FIRST  NODE= ";
780 INPUT N1(I)
790 PRINT "          ''           ";I;"   ''        SECOND NODE= ";
800 INPUT N2(I)
810 PRINT "          ''           ";I;"   ''        MATERIAL... = ";
820 INPUT M3(I)
830 PRINT "          ''           ";I;"   ''        SECTION.... = ";
840 INPUT S2(I)
850 NEXT I
860 PRINT
870 DIM E1(N9),R1(N9)
880 FOR I=1 TO N9
890 PRINT "MATERIAL NU: ";I;" YOUNG'S MODULUS... = ";
900 INPUT E1(I)
910 PRINT "     ''          ";I;" MASS PER UNIT VOL. = ";
920 INPUT R1(I)
930 NEXT I
940 PRINT
950 DIM S1(N8),I1(N8)
960 FOR I=1 TO N8
970 PRINT "SECTION NU: ";I;" AREA... = ";
980 INPUT S1(I)
990 PRINT "    ''          ";I;" INERTIA= ";
1000 INPUT I1(I)
1010 NEXT I
1020 IF R9=0 THEN 1290
1030 DELETE M1,M2,K1
1040 DIM M1(R9,3),M2(R9,3),K1(R9,3)
1050 FOR I=1 TO R9
1060 PRINT "SYSTEM M-S-M NU: ";I;" NUMBER OF FIRST  NODE = ";
1070 INPUT N1(I+P9)
1080 PRINT "MASS       IN X = ";
1090 INPUT M1(I,1)
1100 PRINT " ''         Y = ";
1110 INPUT M1(I,2)
1120 PRINT "INERTIA     PSI: ";
1130 INPUT M1(I,3)
1140 PRINT "SYSTEM M-S-M NU: ";I;" NUMBER OF SECOND NODE = "
1150 INPUT N2(I+P9)
1160 PRINT "MASS       IN X = ";
1170 INPUT M2(I,1)
1180 PRINT " ''         Y = ";
1190 INPUT M2(I,2)
1200 PRINT "INERTIA     PSI= ";
1210 INPUT M2(I,3)
1220 PRINT "STIFFNESS IN X = ";
```

```
1230 INPUT K1(I,1)
1240 PRINT "      ''              Y = ";
1250 INPUT K1(I,2)
1260 PRINT "      ''          PSI= ";
1270 INPUT K1(I,3)
1280 NEXT I
1290 IF Q9=0 THEN 1510
1300 DELETE M4,K4
1310 DIM M4(6,6),K4(6,6)
1320 PRINT
1330 PRINT "ARBITRARY SYSTEM NUMBER OF FIRST  NODE = ";
1340 INPUT N1(P9+R9+1)
1350 PRINT "                      ''            SECOND NODE = ";
1360 INPUT N2(P9+R9+1)
1370 PRINT
1380 FOR I=1 TO 6
1390 FOR J=I TO 6
1400 PRINT "M(";I;",";J;")= ";
1410 INPUT M4(I,J)
1420 NEXT J
1430 NEXT I
1440 PRINT
1450 FOR I=1 TO 6
1460 FOR J=I TO 6
1470 PRINT "K(";I;",";J;")= ";
1480 INPUT K4(I,J)
1490 NEXT J
1500 NEXT I
1510 PRINT
1520 DELETE X,Y
1530 DIM X(N6),Y(N6)
1540 FOR I=1 TO N6
1550 PRINT "NODE  NU: ";I;" ABSCISSA = ";
1560 INPUT X(I)
1570 PRINT "  ''        ";I;" ORDINATE = ";
1580 INPUT Y(I)
1590 NEXT I
1600 PRINT
1610 IF N5=0 THEN 1750
1620 DELETE N4,C1,C2,C3
1630 DIM N4(N5),C1(N5),C2(N5),C3(N5)
1640 FOR I=1 TO N5
1650 PRINT "NUMBER OF ";I;" CONSTRAINED NODE = ";
1660 INPUT N4(I)
1670 PRINT "FOR THIS NODE CONDITION IN U = ";
1680 INPUT C1(I)
1690 PRINT "                      ''          V = ";
1700 INPUT C2(I)
1710 PRINT "                      ''          PSI= ";
1720 INPUT C3(I)
1730 PRINT
1740 NEXT I
1750 REM CONSTRUCTION OF ELEMENTARY MASS AND STIFFNESS MATRICES
1760 REM -------------------------------------------------------
1770 DELETE 100,1760
1780 DELETE K,M,K0,M0
1790 DIM K(N,N),M(N,N),K0(6,6),M0(6,6)
1800 K=0
1810 M=0
1820 P7=0
1830 R7=0
1840 Q7=0
1850 FIND 16
```

```
1860 FOR E=1 TO N7
1870 K9=N0(E)
1880 K0=0
1890 M0=0
1900 GO TO K9 OF 1910,1940,1970
1910 P7=P7+1
1920 P8=P7
1930 GO TO 1980
1940 R7=R7+1
1950 P8=P9+R7
1960 GO TO 1980
1970 P8=P9+R9+1
1980 L1=(N1(P8)-1)*3
1990 L2=(N2(P8)-1)*3
2000 M8=N1(P8)
2010 M9=N2(P8)
2020 X1=X(M8)
2030 X2=X(M9)
2040 Y1=Y(M8)
2050 Y2=Y(M9)
2060 L0=((X2-X1)^2+(Y2-Y1)^2)^0.5
2070 C=(X2-X1)/L0
2080 S=(Y2-Y1)/L0
2090 GO TO K9 OF 2100,2590,2680
2100 REM BEAM MATRICES
2110 M9=M3(P7)
2120 S9=S2(P7)
2130 E0=E1(M9)
2140 R0=R1(M9)
2150 S0=S1(S9)
2160 I0=I1(S9)
2170 A9=E0*S0/L0
2180 A8=E0*I0/L0^3
2190 A1=A9
2200 A2=12*A8
2210 A3=6*L0*A8
2220 A4=4*L0^2*A8
2230 K0(1,1)=A1
2240 K0(1,4)=-A1
2250 K0(4,4)=A1
2260 K0(2,2)=A2
2270 K0(2,3)=A3
2280 K0(2,5)=-A2
2290 K0(2,6)=A3
2300 K0(3,3)=A4
2310 K0(3,5)=-A3
2320 K0(3,6)=A4/2
2330 K0(5,5)=A2
2340 K0(5,6)=-A3
2350 K0(6,6)=A4
2360 B9=R0*S0*L0/6
2370 B8=R0*S0*L0/420
2380 B1=2*B9
2390 B2=156*B8
2400 B3=22*L0*B8
2410 B4=54*B8
2420 B5=-13*L0*B8
2430 B6=4*L0^2*B8
2440 B7=-3*L0^2*B8
2450 M0(1,1)=B1
2460 M0(1,4)=B1/2
2470 M0(4,4)=B1
2480 M0(2,2)=B2
```

```
2490 MO(2,3)=B3
2500 MO(2,5)=B4
2510 MO(2,6)=B5
2520 MO(3,3)=B6
2530 MO(3,5)=-B5
2540 MO(3,6)=B7
2550 MO(5,5)=B2
2560 MO(5,6)=-B3
2570 MO(6,6)=B6
2580 GO TO 2770
2590 REM M-S-M MATRICES
2600 FOR I=1 TO 3
2610 KO(I,I)=K1(R7,I)
2620 KO(I,I+3)=-K1(R7,I)
2630 KO(I+3,I+3)=K1(R7,I)
2640 MO(I,I)=M1(R9,I)
2650 MO(I+3,I+3)=M2(R7,I)
2660 NEXT I
2670 GO TO 2770
2680 REM ARB MATRICES
2690 FOR I=1 TO 6
2700 FOR J=I TO 6
2710 M4(J,I)=M4(I,J)
2720 K4(J,I)=K4(I,J)
2730 NEXT J
2740 NEXT I
2750 MO=M4
2760 KO=K4
2770 REM TRANSFORMATION FROM THE REFERENCE FRAME ATTACHED
2780 REM  TO THE ELEMENT TO THE ABSOLUTE REFERENCE FRAME
2790 FOR I=1 TO 6
2800 FOR J=I TO 6
2810 KO(J,I)=KO(I,J)
2820 MO(J,I)=MO(I,J)
2830 NEXT J
2840 NEXT I
2850 DELETE T1,T2
2860 DIM T1(6,6),T2(6,6)
2870 T1=0
2880 T1(1,1)=C
2890 T1(1,2)=S
2900 T1(2,1)=-S
2910 T1(2,2)=C
2920 T1(3,3)=1
2930 FOR I=1 TO 3
2940 FOR J=1 TO 3
2950 T1(I+3,J+3)=T1(I,J)
2960 NEXT J
2970 NEXT I
2980 T2=KO MPY T1
2990 T1=TRN(T1)
3000 KO=T1 MPY T2
3010 T2=T1 MPY MO
3020 T1=TRN(T1)
3030 MO=T2 MPY T1
3040 WRITE @33:KO,MO
3050 REM POSITIONING INTO GLOBAL MATRICES [M,K]
3060 REM ------------------------------------
3070 FOR I=1 TO 3
3080 FOR J=1 TO 3
3090 M(I+L1,J+L1)=M(I+L1,J+L1)+MO(I,J)
3100 M(I+L1,J+L2)=M(I+L1,J+L2)+MO(I,J+3)
3110 M(I+L2,J+L1)=M(I+L2,J+L1)+MO(I+3,J)
```

```
3120 M(I+L2,J+L2)=M(I+L2,J+L2)+MO(I+3,J+3)
3130 K(I+L1,J+L1)=K(I+L1,J+L1)+KO(I,J)
3140 K(I+L1,J+L2)=K(I+L1,J+L2)+KO(I,J+3)
3150 K(I+L2,J+L1)=K(I+L2,J+L1)+KO(I+3,J)
3160 K(I+L2,J+L2)=K(I+L2,J+L2)+KO(I+3,J+3)
3170 NEXT J
3180 NEXT I
3190 NEXT E
3200 REM APPLICATION OF THE BOUNDARY CONDITIONS
3210 REM ------------------------------------------------------------
3220 DELETE LO
3230 DIM LO(N)
3240 FOR I=1 TO N
3250 LO(I)=I
3260 NEXT I
3270 L1=N
3280 IF N5=0 THEN 3780
3290 FOR I9=1 TO N5
3300 N3=N4(I9)
3310 DELETE C4
3320 DIM C4(3)
3330 C4(1)=C1(I9)
3340 C4(2)=C2(I9)
3350 C4(3)=C3(I9)
3360 FOR I8=1 TO 3
3370 IF C4(I8)=0 THEN 3570
3380 L=(N3-1)*3+I8
3390 FOR L2=1 TO L1
3400 IF LO(L2)=L THEN 3420
3410 NEXT L2
3420 IF L2=L1 THEN 3560
3430 FOR I=1 TO L1
3440 FOR J=L2 TO L1-1
3450 K(I,J)=K(I,J+1)
3460 M(I,J)=M(I,J+1)
3470 NEXT J
3480 NEXT I
3490 FOR I=L2 TO L1-1
3500 LO(I)=LO(I+1)
3510 FOR J=1 TO L1-1
3520 K(I,J)=K(I+1,J)
3530 M(I,J)=M(I+1,J)
3540 NEXT J
3550 NEXT I
3560 L1=L1-1
3570 NEXT I8
3580 NEXT I9
3590 DELETE 100,3580
3600 DELETE KO
3610 DIM KO(L1,L1)
3620 FOR I=1 TO L1
3630 FOR J=1 TO L1
3640 KO(I,J)=K(I,J)
3650 NEXT J
3660 NEXT I
3670 DELETE K
3680 DIM K(L1,L1)
3690 K=KO
3700 FOR I=1 TO L1
3710 FOR J=1 TO L1
3720 KO(I,J)=M(I,J)
3730 NEXT J
3740 NEXT I
```

```
3750 DELETE M
3760 DIM M(L1,L1)
3770 M=KO
3780 PRINT
3790 DELETE B1,B2,B3,B4,B5,B6,B7,T1,T2,M1,M2,M3,M4,A1,A2,A3,A4,KO
3800 DELETE E0,R0,S0,I0,S1,S2,E1,R1,I1,X,Y,X1,X2,Y1,Y2,C1,C2,C3,C4
3810 DELETE A9,A8,K4,N4,M8,M9,S9,C,S
3820 DELETE 3590,3810
3830 FIND 15
3840 APPEND 3870
3850 GOSUB 3870
3860 END
3870 REM

100 REM            PROGRAM NUMBER : 11   * SECOND PART *
110 REM
120 PAGE
130 REM DETERMINATION OF FREQUENCIES AND MODES
140 REM ----------------------------------------------------------
150 N9=N
160 N=L1
170 PRINT
180 DELETE M1,N4
190 PRINT "NUMBER OF MODES REQUIRED        =";
200 INPUT N4
210 DELETE K1,D
220 DIM K1(N,N),D(N,N)
230 PRINT
240 PRINT "ACCURACY ON THE EIGENVECTORS =";
250 INPUT P
260 P=P^2
270 PRINT
280 PRINT "MAXIMUM NUMBER OF ITERATIONS =";
290 INPUT N3
300 PRINT
310 PRINT "ARE THERE RIGID BODY MODES   YES OR NO : ";
320 INPUT A$
330 IF A$="YES" THEN 360
340 IF A$="NO" THEN 390
350 GO TO 300
360 PRINT "IF      K+ALPHA*M        ALPHA = ";
370 INPUT A
380 GO TO 400
390 A=0
400 K8=1
410 PAGE
420 REM
430 REM ITERATIVE PROCEDURE
440 REM ----------------------------------------------------------
450 PRINT
460 FOR I=1 TO N
470 FOR J=1 TO N
480 K1(I,J)=K(I,J)+A*M(I,J)
490 NEXT J
500 NEXT I
510 K1=INV(K1)
520 D=K1 MPY M
530 DELETE K1,V,S1,L
540 DIM V(N,N4),S1(N,N),L(N4)
550 V=0
560 S1=D
570 DELETE X,F
580 DIM X(N),F(N)
```

```
590 W3=0
600 X=1
610 K9=0
620 K9=K9+1
630 F=S1 MPY X
640 F2=0
650 FOR I=1 TO N
660 IF ABS(F(I))-F2<0 THEN 690
670 F2=ABS(F(I))
680 F3=F(I)
690 NEXT I
700 W2=1/F3
710 L(K8)=W2-A
720 X=F/F3
730 IF K9-N3<=0 THEN 860
740 PRINT
750 PRINT "THE SYSTEM DOES NOT CONVERGE AFTER : ";K9;" ITERATIONS"
760 PRINT
770 PRINT "DO YOU WISH TO CHANGE THE NUMBER OF MODES"
780 PRINT "                ''                  THE ACCURACY ON THE EIGENVECTORS"
790 PRINT "                ''                  THE MAXIMUM NUMBER OF ITERATIONS"
800 PRINT "                ''                  ALPHA"
810 PRINT "  YES OR NO : ";
820 INPUT A$
830 IF A$="YES" THEN 170
840 IF A$="NO" THEN 2670
850 GO TO 740
860 IF ABS((W2-W3)/W2)<P THEN 890
870 W3=W2
880 GO TO 620
890 PRINT "MODE NU: ";K8;" CONVERGENCE AFTER : ";K9;
900 PRINT " ITERATIONS"
910 FOR I=1 TO N
920 V(I,K8)=X(I)
930 NEXT I
940 IF K8=>N4 THEN 1320
950 DELETE S1,X,F,T,T1
960 N5=N-K8
970 DIM T(N4,N),T1(N4,N)
980 T=TRN(V)
990 T1=T MPY M
1000 DELETE T,R,R1,R2
1010 DIM R(K8,K8),R1(K8,N5),R2(K8,N5)
1020 FOR I=1 TO K8
1030 FOR J=1 TO N
1040 IF J>K8 THEN 1070
1050 R(I,J)=T1(I,J)
1060 GO TO 1090
1070 J1=J-K8
1080 R1(I,J1)=T1(I,J)
1090 NEXT J
1100 NEXT I
1110 R=INV(R)
1120 R2=R MPY R1
1130 DELETE R,R1,S
1140 DIM S(N,N)
1150 CALL "IDN",S
1160 FOR I=1 TO K8
1170 S(I,I)=0
1180 FOR J=K8+1 TO N
1190 J1=J-K8
1200 S(I,J)=-R2(I,J1)
1210 NEXT J
```

```
1220 NEXT I
1230 DELETE R2
1240 DIM S1(N,N)
1250 S1=D MPY S
1260 DELETE S
1270 K8=K8+1
1280 GO TO 570
1290 REM
1300 REM RESULTS
1310 REM ------------------
1320 PRINT
1330 PRINT "EIGENVALUES"
1340 PRINT
1350 FOR I=1 TO N4
1360 PRINT "L(";I;")= ";L(I)
1370 NEXT I
1380 PRINT
1390 PRINT
1400 PRINT "FREQUENCIES [RAD/SEC]"
1410 PRINT
1420 FOR I=1 TO N4
1430 PRINT "W(";I;")= ";ABS(L(I))^0.5
1440 NEXT I
1450 PRINT
1460 PRINT
1470 PRINT "FREQUENCIES [HERTZ]"
1480 PRINT
1490 FOR I=1 TO N4
1500 PRINT "F(";I;")= ";ABS(L(I))^0.5/2/PI
1510 NEXT I
1520 PRINT
1530 PRINT
1540 PRINT "MODES"
1550 DELETE V1
1560 DIM V1(N9,N4)
1570 K9=1
1580 FOR I=1 TO N9
1590 IF L0(K9)=I THEN 1640
1600 FOR J=1 TO N4
1610 V1(I,J)=0
1620 NEXT J
1630 GO TO 1690
1640 IF K9>N THEN 1600
1650 FOR J=1 TO N4
1660 V1(I,J)=V(K9,J)
1670 NEXT J
1680 K9=K9+1
1690 NEXT I
1700 FOR J=1 TO N4
1710 PRINT
1720 PRINT "MODE NU: ";J
1730 PRINT "------------------"
1740 FOR I=1 TO N6
1750 K9=(I-1)*3
1760 PRINT USING 1770:"NODE  NU: ";I;" U= ";V1(K9+1,J);" V= ";
1770 IMAGE 10A,2D,4A,3E,4A,S
1780 PRINT USING 1790:V1(K9+2,J);" PSI= ";V1(K9+3,J)
1790 IMAGE 3E,6A,3E
1800 NEXT I
1810 NEXT J
1820 DELETE S1,T1,R
1830 DIM S1(N,N4),T1(N4,N),R(N4,N4)
1840 S1=M MPY V
```

```
1850 T1=TRN(V)
1860 R=T1 MPY S1
1870 PRINT
1880 PRINT "MODAL MASS MATRIX"
1890 PRINT R
1900 PRINT
1910 S1=K MPY V
1920 R=T1 MPY S1
1930 PRINT "MODAL STIFFNESS MATRIX"
1940 PRINT R
1950 PRINT
1960 REM
1970 REM ENERGIES OF ELEMENTS
1980 REM --------------------------------
1990 P7=0
2000 R7=0
2010 Q7=0
2020 FIND 16
2030 DELETE K1,M1
2040 DIM K1(N7,N4),M1(N7,N4)
2050 FOR E=1 TO N7
2060 K9=NO(E)
2070 GO TO K9 OF 2080,2110,2140
2080 P7=P7+1
2090 P8=P7
2100 GO TO 2150
2110 R7=R7+1
2120 P8=P9+R7
2130 GO TO 2150
2140 P8=P9+R9+1
2150 L1=(N1(P8)-1)*3
2160 L2=(N2(P8)-1)*3
2170 DELETE S1,T1,R1,K0,M0
2180 DIM S1(6,N4),T1(N4,6),R1(6,N4),K0(6,6),M0(6,6)
2190 FOR I=1 TO 3
2200 I3=I+3
2210 FOR J=1 TO N4
2220 S1(I,J)=V1(L1+I,J)
2230 S1(I3,J)=V1(L2+I,J)
2240 NEXT J
2250 NEXT I
2260 T1=TRN(S1)
2270 READ @33:K0,M0
2280 R1=K0 MPY S1
2290 R=T1 MPY R1
2300 FOR J=1 TO N4
2310 K1(E,J)=R(J,J)
2320 NEXT J
2330 R1=M0 MPY S1
2340 R=T1 MPY R1
2350 FOR J=1 TO N4
2360 M1(E,J)=R(J,J)
2370 NEXT J
2380 NEXT E
2390 DELETE M2,K2
2400 FOR J=1 TO N4
2410 K2=0
2420 M2=0
2430 PRINT
2440 PRINT "MODE NO: ";J
2450 PRINT
2460 FOR E=1 TO N7
2470 K2=K2+K1(E,J)
```

```
2480 M2=M2+M1(E,J)
2490 NEXT E
2500 FOR E=1 TO N7
2510 IF K2<=1.0E-10 THEN 2560
2520 K1(E,J)=K1(E,J)/K2*100
2530 PRINT USING 2540:"ELEM NU: ";E;" STRAIN ENERGY..= ";K1(E,J);" % "
2540 IMAGE 9A,2D,18A,2E,3A
2550 GO TO 2580
2560 PRINT USING 2570:"ELEM NU: ";E;" STRAIN ENERGY..= 0"
2570 IMAGE 9A,2D,19A
2580 NEXT E
2590 PRINT
2600 FOR E=1 TO N7
2610 M1(E,J)=M1(E,J)/M2*100
2620 PRINT USING 2540:"ELEM NU: ";E;" KINETIC ENERGY.= ";M1(E,J);" % "
2630 NEXT E
2640 NEXT J
2650 PRINT
2660 GO TO 760
2670 END

100 REM              PROGRAM NUMBER :12
110 INIT
120 PAGE
130 PRINT "*********************************************"
140 PRINT "*                                           *"
150 PRINT "*        GAUSS-LEGENDRE INTEGRATION         *"
160 PRINT "*          OF A FUNCTION F WHERE :          *"
170 PRINT "*                                           *"
180 PRINT "*   F=F(X)  OR  F=F(X,Y)  OR  F=F(X,Y,Z)    *"
190 PRINT "*                                           *"
200 PRINT "*      THE INTEGRAL IS CALCULATED WITH      *"
210 PRINT "*          2 TO 10 GAUSS POINTS             *"
220 PRINT "*                                           *"
230 PRINT "*********************************************"
240 DELETE 2010,4990
250 A$="BEG"
260 PRINT
270 PRINT "HAS THE FUNCTION 1,2 OR 3 VARIABLES  NV= ";
280 INPUT N2
290 PRINT
300 PRINT "HOW MANY GAUSS POINTS DO YOU WISH    NG= ";
310 INPUT N
320 PRINT
330 IF A$="GAU" THEN 360
340 GO TO 1840
350 IF A$="FUN" THEN 1300
360 N1=N-1
370 DELETE H,A
380 DIM H(10),A(10)
390 GO TO N1 OF 400,430,480,530,600,670,760,850,960
400 A(1)=-0.57735026919
410 H(1)=1
420 GO TO 1060
430 A(1)=-0.774596669241
440 A(2)=0
450 H(1)=0.555555555556
460 H(2)=0.888888888889
470 GO TO 1060
480 A(1)=-0.861136311594
490 A(2)=-0.339981043585
500 H(1)=0.347854845137
510 H(2)=0.652145154863
```

258

```
520 GO TO 1060
530 A(1)=-0.906179845939
540 A(2)=-0.538469310106
550 A(3)=1.0E-12
560 H(1)=0.236926885056
570 H(2)=0.478628670499
580 H(3)=0.568888888889
590 GO TO 1060
600 A(1)=-0.932469514203
610 A(2)=-0.661209386466
620 A(3)=-0.238619186083
630 H(1)=0.171324492379
640 H(2)=0.360761573048
650 H(3)=0.467913934573
660 GO TO 1060
670 A(1)=-0.949107912343
680 A(2)=-0.741531185599
690 A(3)=-0.405845151377
700 A(4)=0
710 H(1)=0.129484966169
720 H(2)=0.279705391489
730 H(3)=0.381830050505
740 H(4)=0.417959183673
750 GO TO 1060
760 A(1)=-0.960289856497
770 A(2)=-0.796666477414
780 A(3)=-0.525532409916
790 A(4)=-0.183434642496
800 H(1)=0.10122853629
810 H(2)=0.222381034453
820 H(3)=0.313706645878
830 H(4)=0.362683783378
840 GO TO 1060
850 A(1)=-0.968160239508
860 A(2)=-0.836031107327
870 A(3)=-0.6133714327
880 A(4)=-0.324253423404
890 A(5)=0
900 H(1)=0.0812743883616
910 H(2)=0.180648160695
920 H(3)=0.260610696403
930 H(4)=0.31234707704
940 H(5)=0.330239355001
950 GO TO 1060
960 A(1)=-0.973906528517
970 A(2)=-0.865063366689
980 A(3)=-0.679409568299
990 A(4)=-0.433395394129
1000 A(5)=-0.148874338982
1010 H(1)=0.0666713443087
1020 H(2)=0.149451349151
1030 H(3)=0.219086362516
1040 H(4)=0.26926671931
1050 H(5)=0.295524224715
1060 FOR I=1 TO INT(N/2)
1070 J=N+1-I
1080 A(J)=-A(I)
1090 H(J)=H(I)
1100 NEXT I
1110 IF A$="GAU" THEN 1300
1120 PRINT
1130 PRINT
1140 PRINT "WHAT IS THE MINIMUM VALUE OF X=.";
```

```
1150 INPUT X1
1160 PRINT "          ''      MAXIMUM   ''       X= ";
1170 INPUT X2
1180 PRINT
1190 IF N2-1=0 THEN 1300
1200 PRINT "WHAT IS THE MINIMUM VALUE OF Y= ";
1210 INPUT Y1
1220 PRINT "          ''      MAXIMUM   ''       Y= ";
1230 INPUT Y2
1240 PRINT
1250 IF N2-2=0 THEN 1300
1260 PRINT "WHAT IS THE MINIMUM VALUE OF Z= ";
1270 INPUT Z1
1280 PRINT "          ''      MAXIMUM   ''       Z= ";
1290 INPUT Z2
1300 I9=0
1310 FOR I=1 TO N
1320 IF N2-1=0 THEN 1360
1330 FOR J=1 TO N
1340 IF N2-2=0 THEN 1360
1350 FOR K=1 TO N
1360 A1=-(X1-X2)/2
1370 B1=(X1+X2)/2
1380 X=A1*A(I)+B1
1390 IF N2-1=0 THEN 1470
1400 A2=-(Y1-Y2)/2
1410 B2=(Y1+Y2)/2
1420 Y=A2*A(J)+B2
1430 IF N2-2=0 THEN 1470
1440 A3=-(Z1-Z2)/2
1450 B3=(Z1+Z2)/2
1460 Z=A3*A(J)+B3
1470 GOSUB 3000
1480 GO TO N2 OF 1490,1510,1530
1490 I9=I9+F*H(I)*A1
1500 GO TO 1560
1510 I9=I9+F*H(I)*H(J)*A1*A2
1520 GO TO 1550
1530 I9=I9+F*H(I)*H(J)*H(K)*A1*A2*A3
1540 NEXT K
1550 NEXT J
1560 NEXT I
1570 PRINT
1580 PRINT
1590 PRINT "   *******************************"
1600 PRINT "   *                             *"
1610 PRINT USING 1620:"   *   INTEGRAL  = ",I9," *"
1620 IMAGE 18A,5 E,3A
1630 PRINT "   *                             *"
1640 PRINT "   *******************************"
1650 PRINT
1660 PRINT
1670 PRINT "DO YOU WISH TO BEGIN AGAIN                         REP=BEG"
1680 PRINT "          ''      CHANGE THE FUNCTION              REP=FUN"
1690 PRINT "          ''      CHANGE THE NUMBER OF GAUSS POINTS REP=GAU"
1700 PRINT "          ''      CHANGE THE LIMITS                REP=LIM"
1710 PRINT "          ''      STOP                             REP=STOP"
1720 PRINT
1730 PRINT "   REP= ";
1740 INPUT A$
1750 IF A$="STOP" THEN 1830
1760 IF A$="LIM" THEN 1120
1770 IF A$="GAU" THEN 290
```

```
1780 IF A$="FUN" THEN 1810
1790 IF A$="BEG" THEN 240
1800 GO TO 1650
1810 DELETE 2010,4990
1820 GO TO 1840
1830 END
1840 PRINT
1850 PRINT
1860 GO TO N2 OF 1870,1890,1910
1870 PRINT "EXPRESSION OF THE FUNCTION      F(X)"
1880 GO TO 1920
1890 PRINT "EXPRESSION OF THE FUNCTION     F(X,Y)"
1900 GO TO 1920
1910 PRINT "EXPRESSION OF THE FUNCTION   F(X,Y,Z)"
1920 PRINT "------------------------------------------------------------"
1930 PRINT
1940 PRINT " PROGRAM THE FUNCTION BEGINNING AT LINE 3000"
1950 PRINT " NUMBER THE LINES IN STEPS OF 10"
1960 PRINT " END THE NUMBERING WITH ''RETURN''"
1970 PRINT " TYPE ON THE KEYBOARD: RUN 5000"
1980 PRINT
1990 END
5000 GO TO 360
```

Appendix: Lagrange's Equations

Lagrange's equations are deduced from the principle of virtual work applied to the forces, including inertia forces, acting on a system. These equations are easy to use and well suited for mechanical systems.

In what follows $q_1, \ldots, q_N$ are a set of N independent generalized coordinates for a system having N degrees of freedom. The N Lagrange's equations which will give the N coupled differential equations of motion are

$$\frac{\mathrm{d}}{\mathrm{d}t}\left(\frac{\partial T}{\partial q_i^\circ}\right) - \frac{\partial T}{\partial q_i} = Q_i \qquad i = 1, \ldots, N \tag{1}$$

with

T, kinetic energy

Q_i, generalized force corresponding to the coordinate q_i.

In most cases, forces induced by strains can be deduced from the potential strain energy U and the forces due to viscous damping from a Rayleigh's dissipation function R. Then (1) becomes

$$\frac{\mathrm{d}}{\mathrm{d}t}\left(\frac{\partial T}{\partial q_i^\circ}\right) - \frac{\partial T}{\partial q_i} + \frac{\partial U}{\partial q_i} + \frac{\partial R}{\partial q_i^\circ} = Q_i \qquad i = 1, \ldots, N \tag{2}$$

In (2), Q_i is the vector of external forces. This expression is commonly used in this book.

When the generalized coordinates are no longer independent it is necessary to eliminate some of the coordinates in T, U, R before applying (1) or (2). If this elimination cannot be performed, Lagrange multipliers have to be used. In this book the generalized coordinates are always independent.

Bibliography

1. Bathe, K. J. and Wilson, E., *Numerical Methods in Finite Element Analysis*, Prentice-Hall (1976).
2. Bishop, R. E. D., Gladwell, G. M. L., and Michaelson, S., *The Matrix Analysis of Structures*, Cambridge University Press (1965).
3. Clough, R. W., and Penzien, J., *Dynamics of Structures*, McGraw-Hill (1975).
4. Craig, R. R., *Structural Dynamics*, Wiley (1981).
5. Den Hartog, J. P., *Mechanical Vibrations*, 3rd edition, McGraw-Hill (1947).
6. Dimaragonas, A. D., *Vibration Engineering*, West Publishing (1976).
7. Harris, C. M., and Crede, C. E., *Shock and Vibration Handbook*, 2nd edition, McGraw-Hill (1976).
8. Levy, S., and Wilkinson, J. P. D., *The Component Element Method in Dynamics*, McGraw-Hill (1976).
9. McCallion, H., *Vibration of Linear Mechanical Systems*, William Clover (1973).
10. Marguerre, K., and Wölfel, K., *Mechanics of Vibration*, Sijthoff and Noordhoff (1979).
11. Meirovitch, L., *Elements of Vibration Analysis*, McGraw-Hill (1975).
12. Seto, W. W., *Theory and Problems of Mechanical Vibrations*, Schaum (1964).
13. Thomson, W. T., *Theory of Vibration with Applications*, 2nd edition, Prentice-Hall (1981).
14. Timoshenko, S. P., Young, D. H., and Weaver, W., *Vibration Problems in Engineering*, 4th edition, Wiley (1974).
15. Tse, F. S., Morse, I. E., and Hinkle, R. T., *Mechanical Vibrations. Theory and Applications*, 2nd edition, Allyn and Bacon (1978).
16. Warburton, G. B., *The Dynamical Behaviour of Structures*, 2nd edition, Pergamon (1976).

The above bibliography presents the existing state of knowledge in mechanical vibrations for engineers; advances in this knowledge are published in a wide variety of research journals. A comprehensive set of abstracts from these journals as well as surveys of the latest technology in shock and vibration can be found in *The Shock and Vibration Digest* which is available from the Shock and Vibration Information Center, Naval Research Laboratory, Washington, DC 20375, U.S.A.

Index